Programming & Analysis (PA)

ARE 5.0 Exam Guide
(Architect Registration Examination)

ARE 5.0 Overview, Exam Prep Tips, Guide, and Critical Content

Gang Chen

ArchiteG®, Inc.
Irvine, California

Programming & Analysis (PA) ARE 5.0 Exam Guide (Architect Registration Examination): ARE 5.0 Overview, Exam Prep Tips, Guide, and Critical Content

V15

Copy Editor: Penny L Kortje

ArchiteG®, Inc.
http://www.ArchiteG.com

ISBN: 9781612650487

PRINTED IN THE UNITED STATES OF AMERICA

What others are saying about *ARE Mock Exam series* & *ARE Exam Guide series* ...

"Great study guide..."
"This was a great resource supplement to my other study resources. I appreciated the mock exam questions the most, and the solutions offer an explanation as to why the answer is correct. I will definitely check out his other ARE exam resources!

UPDATE: Got my PASS Letter!"
—Sean Primeaux

"Tried everything 4 times before reading this book and PASSED!"
"I had failed this exam 4 times prior to getting this book…I had zero clue as to what I was doing wrong. I read Ballast, Kaplan and random things on the forum but for the life of me couldn't pinpoint where I was missing it until I read THIS BOOK! Gang did an excellent job…I remember … reading Gang's book and saying Ohhhh like 4 or 5 times. I read his book several times until I became comfortable with the information. I went in on test day and it was a breeze. I remember walking out of there thinking I couldn't believe I struggled so much before. The tips in here are priceless! I strongly recommend this book…"
—hendea1

"Add this to your ARE study"
"This was a very helpful practice exam and discussion. I really appreciated the step-by-step review of the author's approach... As I studied it last before taking the test, Gang Chen's book probably made the difference for me."
—Dan Clowes ("XLine")

"Good supplemental mock exam"
"I found the mock exam to be very helpful, all of the answers are explained thoroughly and really help you understand why it is correct...Also the introduction and test taking tips are very helpful for new candidates just starting the ARE process."
—Bgrueb01

"Essential Study Tool"
"I have read the book and found it to be a great study guide for myself. Mr. Gang Chen does such a great job of helping you get into the right frame of mind for the content of the exam. Mr. Chen breaks down the points on what should be studied and how to improve your chances of a pass with his knowledge and tips for the exam.

I highly recommend this book to anyone…it is an invaluable tool in the preparation for the exam as Mr. Chen provides a vast amount of knowledge in a very clear, concise, and logical matter."
—Luke Giaccio

"Wish I had this book earlier"
"...The questions are written like the NCARB questions, with various types...check all that apply, fill in the blank, best answer, etc. The answer key helpfully describes why the correct answer is correct, and why the incorrect answers are not. Take it from my experience, at half the cost of other mock exams, this is a must buy if you want to pass..."
—Domiane Forte ("Vitruvian Duck")

"This book did exactly like the others said."
"This book did exactly like the others said. It is immensely helpful with the explanation... There are so many codes to incorporate, but Chen simplifies it into a methodical process. Bought it and just found out I passed. I would recommend."
—Dustin

"It was the reason I passed."
"This book was a huge help. I passed the AREs recently and I felt this book gave me really good explanations for each answer. It was the reason I passed."
—Amazon Customer

"Great Practice Exam"
"... For me, it was difficult to not be overwhelmed by the amount of content covered by the Exam. This Mock Exam is the perfect tool to keep you focused on the content that matters and to evaluate what you know and what you need to study. It definitely helped me pass the exam!!"
—Michael Harvey ("Harv")

"One of the best practice exams"
"Excellent study guide with study tips, general test info, and recommended study resources. Hands down one of the best practice exams that I have come across for this exam. Most importantly, the practice exam includes in depth explanations of answers. Definitely recommended."
—Taylor Cupp

"Great Supplement!!"
"This publication was very helpful in my preparation for my BS exam. It contained a mock exam, followed by the answers and brief explanations to the answers. I would recommend this as an additional study material for this exam."
—Cynthia Zorrilla-Canteros ("czcante")

"Fantastic! "
"When I first began to prepare for this exam; the number of content areas seemed overwhelming and daunting at best. However, this guide clearly dissected each content area into small management components. Of all the study guides currently available for this test - this exam not only included numerous resources (web links, you tube clips, etc..), but also the sample test was extremely helpful. The sample test incorporated a nice balance of diagrams, calculations and general concepts - this book allowed me to highlight any "weak" content areas I had prior to the real exam. In short - this is an awesome book!"
—Rachel Casey (RC)

Dedication

To my parents, Zhuixian and Yugen,
my wife Yongqing Fang,
and my daughters,
Alice, Angela, Amy, and Athena.

Disclaimer

Programming & Analysis (PA) ARE 5.0 Exam Guide (Architect Registration Examination) provides general information about Architect Registration Examination. The book is sold with the understanding that neither the publisher nor the authors are providing legal, accounting, or other professional services. If legal, accounting, or other professional services are required, seek the assistance of a competent professional firm.

The purpose of this publication is not to reprint the content of all other available texts on the subject. You are urged to read other materials, and tailor them to fit your needs.

Great effort has been taken to make this resource as complete and accurate as possible. However, nobody is perfect and there may be typographical errors or other mistakes present. You should use this book as a general guide and not as the ultimate source on this subject. If you find any potential errors, please send an email to:
info@ArchiteG.com

Programming & Analysis (PA) ARE 5.0 Exam Guide (Architect Registration Examination) is intended to provide general, entertaining, informative, educational, and enlightening content. Neither the publisher nor the author shall be liable to anyone or any entity for any loss or damages, or alleged loss or damages, caused directly or indirectly by the content of this book.

If you do not wish to be bound by the above, you may return this book to the publisher for a full refund.

Legal Notice

ARE Exam Guide series by ArchiteG, Inc.

Time and effort are the most valuable asset of a candidate. How to cherish and effectively use your limited time and effort is the key of passing any exam. That is why we publish the ARE Exam Guide series and the ARE Mock Exam series to help you to study and pass the ARE exams in the shortest time possible. We have done the hard work so that you can save time and money. We do not want to make you work harder than you have to. To save your time, we use a *standard* format for all our ARE 5.0 Exam Guide books and ARE 5.0 Mock Exam books, so that you can quickly skip the *identical* information you have already read in other books of the series and go straight to the *unique* "meat and potatoes" portion of the book.

The trick and the most difficult part of writing a good book is to turn something that is very complicated into something that is very simple. This involves researching and really understanding very complicated materials, absorbing the information, and then writing about the topic in a way that makes it very easy to understand. To succeed at this, you need to know the materials very well. Our goal is to write books that are clear, concise, and helpful.

Do not force yourself to memorize a lot of numbers. Read through the numbers a few times, and you should have a very good impression of them. You need to make the judgment call: If you miss a few numbers, you can still pass the exam, but if you spend too much time drilling these numbers, you may miss out on the big pictures and fail the exam.

The existing ARE prep books by others are either way too easy or way over-killed. They do NOT match the real ARE 5.0 exams at all.

We have done very comprehensive research on the official NCARB guides, many related websites, reference materials, and other available ARE exam prep materials. We match our mock exams as close as possible to the NCARB samples and the real ARE exams. Some readers had failed an ARE exam two or three times before, and they eventually passed the exam with our help.

All our mock exam books include a complete set of questions and case studies. We try to mimic the real ARE exams by including the same number of questions, using a similar format, and asking the same type of questions. We also include detailed answers and explanations to our questions.

There is some extra information on ARE overviews and exam-taking tips in Chapter One. This is based on NCARB *and* other valuable sources. This is a bonus feature we included in each book because we want our readers to be able to buy our ARE mock exam books together or individually. We want you to find all necessary ARE exam information and resources at one place and through our books.

All our books are available at
http://www.GreenExamEducation.com

How to Use Our Books

We suggest you read *Programming & Analysis (PA) ARE 5.0 Exam Guide (Architect Registration Examination)* at least three times:

Read once and cover all chapters, the Appendixes, the related *free* PDF files, and other resources. Highlight the information you are not familiar with.

Read twice focusing on the highlighted information to memorize. You can repeat this process as many times as you want until you master the content of the book.

After reviewing these materials, you can read the related mock exam book and take the mock exam, and then check your answers against the answers and explanations in the back, including explanations for the questions you answer correctly. You may have answered some questions correctly for the wrong reason. Highlight the information you are not familiar with.

Like the real exam, the mock exam will continue to use **multiple choice, check-all-that-apply,** and **quantitative fill-in-the-blank**. There are also three new question types: **hotspots, case studies,** and **drag-and-place**.

Review your highlighted information in both our ARE guidebook and mock exam book and take the mock exam again. Try to answer 100% of the questions correctly this time. Repeat the process until you can answer all the questions correctly.

Our books include most if not all the information you need to do the calculations, as well as step-by-step explanations. After reading our books, you will greatly improve your ability to deal with the real ARE questions and have a great chance of passing the exam on the first try.

Read our related mock exam book and take the mock exam at least two weeks before the real exam. You should *not* wait until the night before the real exam to take the mock exam. If you do not do well, you will go into panic mode and NOT have enough time to review your weaknesses.

Read for the final time the night before the real exam. Review *only* the information you highlighted, especially the questions you did not answer correctly when you took the mock exam for the first time.

Our books are very light so you can easily carry them around. These features will allow you to review our books whenever you have a few minutes.

The Table of Contents is very detailed so you can locate information quickly. If you are on a tight schedule you can forgo reading our books linearly and jump around to the sections you need.

Table of Contents

Appendixes

Back Page Promotion

Index

Chapter One

Overview of the Architect Registration Examination (ARE)

A. First Thing First: Go to the Website of your Architect Registration Board and Read all the Requirements of Obtaining an Architect License in your Jurisdiction
See the following link:
https://www.ncarb.org/get-licensed/state-licensing-boards

B. Download and Review the Latest ARE Documents at the NCARB Website

1. Important links to the FREE and official NCARB documents
NCARB launched ARE 5.0 on November 1, 2016. The current version of the Architect Registration Examination (ARE 5.0) includes six divisions:

- Practice Management (PcM)
- Project Management (PjM)
- Programming & Analysis (PA)
- Project Planning & Design (PPD)
- Project Development & Documentation (PDD)
- Construction & Evaluation (CE)

All ARE divisions continue to use **multiple choice, check-all-that-apply,** and **quantitative fill-in-the-blank**. The new exams include three new question types: **Hotspots, case studies,** and **drag-and-place**.

There is a tremendous amount of valuable information covering every step of becoming an architect available free of charge at the NCARB website:
http://www.ncarb.org/

For example, you can find guidance about architectural degree programs accredited by the National Architectural Accrediting Board (NAAB), NCARB's Architectural Experience Program (AXP) formerly known as Intern Development Program (IDP), and licensing requirements by state. These documents explain how you can qualify to take the Architect Registration Examination.

We find the official ARE 5.0 Guidelines, ARE 5.0 Handbook, and ARE 5.0 Credit Model extremely valuable. See the following link:
http://www.ncarb.org/ARE/ARE5.aspx

You should start by studying these documents.

2. **A detailed list and brief description of the FREE PDF files that you can download from NCARB**
The following is a detailed list of the FREE PDF files that you can download from NCARB. They are listed in order based on their importance.

    - All **ARE 5.0** information can be found at the following links**:**
    http://www.ncarb.org/ARE/ARE5.aspx
    https://www.ncarb.org/blog/are-50-study-materials

    - **ARE 5.0 Guidelines** includes extremely valuable information on the ARE overview, NCARB, registration (licensure), architectural education requirements, the Architectural Experience Program (AXP), establishing your eligibility to test, scheduling an exam appointment, taking the ARE, receiving your score, retaking the ARE, the exam format, scheduling, and links to other FREE NCARB PDF files. You need to read this at least twice.

    - **NCARB Education Guidelines** contains information on education requirements for initial licensure and for NCARB certification, satisfying the education requirement, foreign-educated applicants, the education alternative to NCARB certification, the Education Evaluation Services for Architects (EESA), the Education Standard, and other resources.

    - **Architectural Experience Program (AXP) Guidelines** includes information on AXP overview, getting started and creating your NCARB record, experience areas and tasks, documenting your experience through hours, documenting your experience through a portfolio, and the next steps. You need to read this document at least twice. The information is valuable.

      NCARB renamed the **Intern Development Program (IDP)** as **Architectural Experience Program (AXP)** in June 2016. Most of NCARB's 54-member boards have adopted the AXP as a prerequisite for initial architect licensure. Therefore, you should be familiar with the program.

      The AXP application fee is $100. This fee includes one free transmittal of your Record for initial registration and keeps your Record active for the first year. After the initial year, there is an annual renewal fee required to maintain an active Record until you become registered. The cost is currently $85 each year. The fees are subject to change, and you need to check the NCARB website for the latest information.

      There are two ways to meet the AXP requirements. The **first method** is **reporting hours**. Most candidates will use this method. You will need to document at least 3,740 required hours under the six different experience areas to complete the program. A minimum of 50% of your experience must be completed under the supervision of a qualified architect.

The following chart lists the hours required under the six experience areas:

Experience Area	Hours Required
Practice Management	160
Project Management	360
Programming & Analysis	260
Project Planning & Design	1,080
Project Development & Documentation	1,520
Construction & Evaluation	360
Total	**3,740**

Figure 1.1 The hours required under the six experience areas

Your experience reports will fall under one of **two experience settings**:
• **Setting A**: Work performed for an architecture firm.
• **Setting O**: Experiences performed outside an architecture firm.

You must earn at least **1,860 hours** in experience **setting A**.

Your AXP experience should be reported to NCARB at least every *six* months and logged within two months of completing each reporting period (the **Six-Month Rule**).

The **second method** to meet AXP requirements is to create an **AXP Portfolio**. This new method is for experienced design professionals who put their licensure on hold and allows you to prove your experience through the preparation of an online portfolio.

To complete the AXP through the **second method**, you will need to meet ALL the AXP criteria through the portfolio. In other words, you cannot complete the experience requirement through a combination of the **AXP portfolio** and **reporting hours**.

See the following link for more information on AXP:
https://www.ncarb.org/gain-axp-experience

- **Certification Guidelines** by NCARB (Skimming through this should be adequate. You should also forward a copy of this PDF file to your AXP supervisor.)

  See the following link which contains resources for supervisors and mentors:
  http://www.ncarb.org/Experience-Through-Internships/Supervisors-and-Mentors/Resources-for-Supervisor-and-Mentors.aspx

- **ARE 5.0 Related FAQs (Frequently Asked Questions)**: Skimming through this should be adequate.

- **Your Guide to ARE 5.0** includes information on understanding the basics of ARE 5.0, new question types, taking the test, making the transition, getting ARE 5.0 done, and planning your budget. The document also contains FAQs, and links for more information. You need to read this document *at least twice*. The information is valuable.

- **ARE 5.0 Handbook** contains an ARE overview, detailed information for each ARE division, and ARE 5.0 references. This handbook explains what NCARB expects you to know so that you can pass the ARE exams. ARE 5.0 uses either **Understand/Apply (U/A)** or **Analyze/Evaluate (A/E)** to designate the appropriate cognitive complexity of each objective, but *avoids* the use of **"Remember,"** the lowest level of cognitive complexity (CC), or **"Create,"** the highest level of CC.

  This handbook has some sample questions for each division. The real exam is like the samples in this handbook.

  Tips:
  - ***ARE 5.0 Handbook*** *has about 180 pages. To save time, you can just read the generic information at the front and back portion of the handbook, and focus on the ARE division(s) you are currently studying for. As you progress in your testing, you can read the applicable division that you are studying for. This way, the content will always be fresh in your mind.*
  - *You need to read this document at least three times. The information is valuable.*

- **ARE 5.0 Test Specification** identifies the ARE 5.0 division structure and defines the major content areas, called **Sections**; the measurement **Objectives**; and the percentage of content coverage, called **Weightings**. This document specifies the scope and objectives of each ARE division, and the percentage of questions in each content category. You need to read this document *at least twice*. The information is valuable, and is the base of all ARE exam questions.

- **ARE 5.0 Prep Videos** include one short video for each division. These videos give you a very good basic introduction to each division, including sample questions and answers, and explanations. You need to watch each video *at least three times*. See the following link:
  https://www.ncarb.org/blog/are-50-study-materials

- **The Burning Question: Why do we need an ARE anyway?** (We do not want to give out a link for this document because it is too long and keeps changing. You can Google it with its title. Skimming through this document should be adequate.)

- **Defining Your Moral Compass** (You can Google it with its title plus the word "NCARB." Skimming through this document should be adequate.)

- **Rules of Conduct** is available as a *free* PDF file at:
  http://www.ncarb.org/
  (Skimming through this should be adequate.)

C. The Intern Development Program (IDP)/Architectural Experience Program (AXP)

1. What is IDP? What is AXP?

IDP is a comprehensive training program jointly developed by the National Council of Architectural Registration Boards (NCARB) and the American Institute of Architects (AIA) to ensure that interns obtain the necessary skills and knowledge to practice architecture independently. NCARB renamed the **Intern Development Program (IDP)** as **Architectural Experience Program (AXP)** in June 2016.

2. Who qualifies as an intern?

Per NCARB, if an individual meets one of the following criteria, s/he qualifies as an intern:
a. Graduates from NAAB-accredited programs
b. Architecture students who acquire acceptable training prior to graduation
c. Other qualified individuals identified by a registration board

D. Overview of the Architect Registration Examination (ARE)

1. How to qualify for the ARE?

A candidate needs to qualify for the ARE via one of NCARB's member registration boards, or one of the Canadian provincial architectural associations.

Check with your Board of Architecture for specific requirements.

For example, in California, a candidate must provide verification of a minimum of <u>five</u> years of education and/or architectural work experience to qualify for the ARE.

Candidates can satisfy the five-year requirement in a variety of ways:

- Provide verification of a professional degree in architecture through a program that is accredited by NAAB or CACB.

  OR
- Provide verification of at least five years of educational equivalents.

  OR
- Provide proof of work experience under the direct supervision of a licensed architect.

See the following link:
http://www.ncarb.org/ARE/Getting-Started-With-the-ARE/Ready-to-Take-the-ARE-Early.aspx

2. How to qualify for an architect license?

Again, each jurisdiction has its own requirements. An individual typically needs a combination of about *eight* years of education and experience, as well as passing scores on the ARE exams. See the following link:
http://www.ncarb.org/Reg-Board-Requirements

For example, the requirements to become a licensed architect in California are:

- Eight years of post-secondary education and/or work experience as evaluated by the Board (including at least one year of work experience under the direct supervision of an architect licensed in a U.S. jurisdiction or two years of work experience under the direct supervision of an architect registered in a Canadian province)
- Completion of the Architectural Experience Program (AXP)
- Successful completion of the Architect Registration Examination (ARE)
- Successful completion of the California Supplemental Examination (CSE)

California does NOT require an accredited degree in architecture for examination and licensure. However, many other states do.

3. **What is the purpose of ARE?**
The purpose of ARE is NOT to test a candidate's competency on every aspect of architectural practice. Its purpose is to test a candidate's competency on providing professional services to protect the *health, safety, and welfare* of the public. It tests candidates on the *fundamental* knowledge of pre-design, site design, building design, building systems, and construction documents and services.

The ARE tests a candidate's competency as a "specialist" on architectural subjects. It also tests her abilities as a "generalist" to coordinate other consultants' works.

You can download the exam content and references for each of the ARE divisions at the following link:
https://www.ncarb.org/pass-the-are/start

4. **What is NCARB's rolling clock?**
   a. Starting on January 1, 2006, a candidate MUST pass ALL ARE sections within five years. A passing score for an ARE division is only valid for five years, and a candidate has to retake this division if she has NOT passed all divisions within the five-year period.

   b. Starting on January 1, 2011, a candidate who is authorized to take ARE exams MUST take at least one division of the ARE exams within five years of the authorization. Otherwise, the candidate MUST apply for the authorization to take ARE exams from an NCARB member board again.

These rules were created by the **NCARB's rolling clock** resolution and passed by NCARB council during the 2004 NCARB Annual Meeting.

5. **How to register for an ARE exam?**
See the instructions in the new ARE guideline at the following link:
http://www.ncarb.org/ARE/ARE5.aspx

6. **How early do I need to arrive at the test center?**
Be at the test center at least 30 minutes BEFORE your scheduled test time, OR you may lose your exam fee.

7. **Exam format & time**
All ARE divisions are administered and graded by computer. The time allowances are as follows:

Division	Number of Questions	Test Duration	Appointment Time
Practice Management	80	2 hr 45 min	3 hr 30 min
Project Management	95	3 hr 15 min	4 hr
Programming & Analysis	95	3 hr 15 min	4 hr
Project Planning & Design	120	4 hr 15 min	5 hr
Project Development & Documentation	120	4 hr 15 min	5 hr
Construction & Evaluation	95	3 hr 15 min	4 hr
Total Time:		21 hr	25 hr 30 min

Figure 1.2 Exam format & time

Remote proctoring will be introduced mid December 2020. After December 13, 2020, the number of questions and time allotted will change to accommodate remote proctoring:

Division	Number of Questions	Test Duration	Appointment Time
Practice Management	65	2 hr 40 min	3 hr 20 min
Project Management	75	3 hr	3 hr 40 min
Programming & Analysis	75	3 hr	3 hr 40 min
Project Planning & Design	100	4 hr 5 min	5 hr
Project Development & Documentation	100	4 hr 5 min	5 hr
Construction & Evaluation	75	3 hr	3 hr 40 min
Total Time:		19 hr 50 min	24 hr 20 min

Figure 1.3 New Exam format & time

NCARB suggests you arrive at the test center a minimum of 30 minutes before your scheduled appointment. You can have one flexible 15-minute break for each division. That is why the appointment time is 45 minutes longer than the actual test time for each division.

Practice Management has 80 questions and NCARB allows you 2 hours and 45 minutes to complete the exam, so you should spend an average of (2x60+45)/80=165/80= 2.06 minutes on each question.

Project Management and **Programming & Analysis** as well as **Construction & Evaluation** each have 95 questions and NCARB allows you 3 hours and 15 minutes to complete each exam, so you should spend an average of (3x60+15)/80=195/95= 2.05 minutes on each question.

Project Planning & Design as well as **Project Development & Documentation** each have 120 questions and NCARB allows you 4 hours and 15 minutes to complete each exam, so you should spend an average of (4x60+15)/120=255/120= 2.13 minutes on each question.

To simplify this, we suggest you spend about 2 minutes for each question in ALL divisions.

8. **How are ARE scores reported?**
   All ARE scores are reported as Pass or Fail. ARE scores are typically posted within seven to ten days. See the instructions in the new ARE guideline at the following link:
   http://www.ncarb.org/ARE/ARE5.aspx

9. **Is there a fixed percentage of candidates who pass the ARE exams?**
   No, there is *not* a fixed percentage of passing or failing. If you meet the minimum competency required to practice as an architect, you pass. The passing scores are the same for all Boards of Architecture.

10. **When can I retake a failed ARE division?**
    You can retake a failed division of the ARE sixty days after the previous attempt. You can only take the same ARE division three times within any twelve-month period.

11. **How much time do I need to prepare for each ARE division?**
    Every person is different, but on average you need about forty to eighty hours to prepare for each ARE division. You need to set a realistic study schedule and stick with it. Make sure you allow time for personal and recreational commitments. If you are working full time, my suggestion is that you allow no less than two weeks but NOT more than two months to prepare for each ARE division. You should NOT drag out the exam prep process too long and risk losing your momentum.

12. **Which ARE division should I take first?**
    This is a matter of personal preference, and you should make the final decision.

    Some people like to start with the easier divisions and pass them first. This way, they build more confidence as they study and pass each division.

    Other people like to start with the more difficult divisions so that if they fail, they can keep busy studying and taking the other divisions while the clock is ticking. Before they know it, six months has passed, and they can reschedule if need be.

13. **ARE exam prep and test-taking tips**
    You can start with Construction & Evaluation (CE) because it gives a limited scope, and you may want to study building regulations and architectural history (especially famous

architects and buildings that set the trends at critical turning points) before you take other divisions.

Complete mock exams and practice questions and case studies, including those provided by NCARB's practice program and this book, to hone your skills.

Form study groups and learn the exam experience of other ARE candidates. The forum at our website is a helpful resource. See the following links:
http://GreenExamEducation.com/
http://GeeForum.com/

Take the ARE exams as soon as you become eligible, since you probably still remember portions of what you learned in architectural school, especially structural and architectural history. Do not make excuses for yourself and put off the exams.

The following test-taking tips may help you:

- Pace yourself properly. You should spend about two minutes for each question on average.
- Read the questions carefully and pay attention to words like *best, could, not, always, never, seldom, may, false, except,* etc.
- For questions that you are not sure of, eliminate the obvious wrong answer and then make an educated guess. Please note that if you do NOT answer the question, you automatically lose the point. If you guess, you at least have a chance of getting it right.
- If you have no idea what the correct answer is and cannot eliminate any obvious wrong answers, then do not waste too much time on the question and just guess. Try to use the same guess answer for all of the questions you have no idea about. For example, if you choose "d" as the guess answer, then you should be consistent and use "d" whenever you have no clue. This way, you are likely having a better chance at guessing more answers correctly.
- Mark the difficult questions, answer them, and come back to review them AFTER you finish all questions. If you are still not sure, go with your first choice. Your first choice is often the best choice.
- You really need to spend time practicing to become VERY familiar with NCARB's question types. This is because ARE is a timed test, and you do NOT have time to learn about the question types during the test. If you do not know them well, you will NOT be able to finish your solution on time.
- The ARE exams test a candidate's competency to provide professional services protecting the *health, safety, and welfare* of the public. Do NOT waste time on aesthetic or other design elements not required by the program.

ARE exams are difficult, but if you study hard and prepare well, combined with your experience, AXP training, and/or college education, you should be able to pass all divisions and eventually be able to call yourself an architect.

14. Strategies for passing ARE exams on the first try

Passing ARE exams on the first try, like everything else, needs not only hard work, but also great strategy.

- **Find out how much you already know and what you should study**

  You goal is NOT to read all the study materials. Your goal is to pass the exam. Flip through the study materials. If you already know the information, skip these parts.

  Complete the NCARB sample questions for the ARE exam you are preparing for NOW without ANY studying. See what percentage you get right. If you get 71% right, you should be able to pass the real exam without any studying. If you get 50% right, then you just need 21% more to pass.

  This "truth-finding" exam or exercise will also help you to find out what your weakness areas are, and what to focus on.

  Look at the same questions again at the end of your exam prep and check the differences.

  Note:
  We suggest you study the sample questions in the official NCARB Study Guide first, and then other study materials, and then come back to NCARB sample questions again several days before the real ARE exam.

  Per NCARB, with the launch of the updated Architect Registration Examination (ARE) 5.0 in December 2020, the new cutting scores are based on the following information:

  "**How Many Questions Do I Need Correct to Pass?**

  Each division of the ARE measures different content knowledge areas. The difference in knowledge areas and the relative difficulty of the questions that make up that content area vary between divisions; therefore, expectations around how many questions you will need to answer correctly also changes from division to division.

  - **Project Development & Documentation and Construction & Evaluation** require the lowest percentage of scored items to be answered correctly to pass. You need to answer between **57 – 62 percent** of scored items correctly on these divisions to pass.
  - **Practice Management and Project Management** require a slightly higher percentage of questions to be answered correctly to pass. You need to answer between **62 – 68 percent** of scored items correctly on these divisions to pass.
  - **Programming & Analysis and Project Planning & Design** require the highest percentage of questions to be answered correctly to pass. You need to answer between **65 – 71 percent** of scored items correctly on these divisions to pass."

For detailed information, see the following link:
https://www.ncarb.org/blog/what-score-do-you-need-to-pass-the-are

- **Cherish and effectively use your limited time and effort**

  Let me paraphrase a story.
  One time someone had a chance to talk with Napoleon. He said:
  "You are such a great leader and have won so many battles, that you can use one of your soldiers to defeat ten enemy soldiers."

  Napoleon responded:
  "That may be true, but I always try to create opportunities where ten of my soldiers fight one enemy soldier. That is why I have won so many battles."

  Whether this story is true is irrelevant. The important thing that you need to know is **how to concentrate your limited time and effort to achieve your goal. Do NOT spread yourself too thin**. This is a principle many great leaders know and use and is why great leaders can use ordinary people to achieve extraordinary goals.

  Time and effort are the most valuable asset of a candidate. How to cherish and effectively use your limited time and effort is the key to passing any exam.

  If you study very hard and read many books, you are probably wasting your time. You are much better off picking one or two good books, covering the major framework of your exams, and then doing two sets of mock exams to find your weaknesses. You WILL pass if you follow this advice. You may still have minor weakness, but you will have covered your major bases.

- **Do NOT stretch your exam prep process too long**
  If you do this, it will hurt instead of helping you. You may forget the information by the time you take the exam.

  Spend 40 to 80 hours for each division (a maximum of two months for the most difficult exams if you really need more time) should be enough. Once you decide on taking an exam, put in 100% of your effort and read the RIGHT materials. Allocate your time and effort on the most important materials, and you will pass.

- **Resist the temptation to read too many books and limit your time and effort to read only a few selected books or a few sections of each book in detail**
  Having all the books but not reading them, or digesting ALL the information in them will not help you. It is like someone having a garage full of foods, and not eating or eating too much of them. Neither way will help.

  You can only eat three meals a day. Similarly, you can ONLY absorb a certain amount of information during your exam prep. So, focus on the most important stuff.

Focus on your weaknesses but still read the other info. The key is to understand, digest the materials, and retain the information.

It is NOT how much you have read, but how much you understand, digest, and retain that counts.

The key to passing an ARE exam, or any other exam, is to know the scope of the exam, and not to read too many books. Select one or two really good books and focus on them. Understand the content and memorize it. For your convenience, I have underlined the fundamental information that I think is very important. You definitely need to *memorize* all the information that I have underlined.

You should try to understand the content first, and then memorize the content of the book by reading it multiple times. This is a much better way than relying on "mechanical" memory without understanding.

When you read the materials, always keep the following in mind:

- **Think like an architect.**
  For example, when you take the Project Development & Documentation (PDD) exam, focus on what need to know to be able to coordinate your engineer's work, or tell them what to do. You are NOT taking an exam for becoming a structural engineer; you are taking an exam to become an architect.

  This criterion will help you filter out the materials that are irrelevant and focus on the right information. You will know what to flip through, what to read line by line, and what to read multiple times.

  I have said this one thousand times, and I will say it once more:
  Time and effort is the most valuable asset of a candidate. How to cherish and effectively use your limited time and effort is the key to passing any exam.

15. **ARE exam preparation requires short-term memory**
    You should understand that ARE Exam Preparation requires **Short-Term Memory**. This is especially true for the MC portion of the exam. You should schedule your time accordingly: in the early stages of ARE exam Preparation, you should focus on understanding and an **initial** review of the material; in the late stages of exam preparation, you should focus on memorizing the material as a **final** review.

16. **Allocation of your time and scheduling**
    You should spend about 60% of your effort on the most important and fundamental study materials, about 30% of your effort on mock exams, and the remaining 10% on improving your weakest areas, i.e., reading and reviewing the questions that you answered incorrectly, reinforcing the portions that you have a hard time memorizing, etc.

    Do NOT spend too much time looking for obscure ARE information because NCARB will HAVE to test you on the most **common** architectural knowledge and information. At least 80% to 90% of the exam content will have to be the most common, important and fundamental knowledge. The exam writers can word their questions to be tricky or confusing, but they have to limit themselves to the important content; otherwise, their tests will NOT be legally defensible. At most, 10% of their test content can be obscure information. You only need to answer about 71% of all the questions correctly. So, if you master the common ARE knowledge (applicable to 90% of the questions) and use the guess technique for the remaining 10% of the questions on the obscure ARE content, you will do well and pass the exam.

On the other hand, if you focus on the obscure ARE knowledge, you may answer the entire 10% obscure portion of the exam correctly, but only answer half of the remaining 90% of the common ARE knowledge questions correctly, and you will fail the exam. We have seen many smart people who can answer very difficult ARE questions correctly because they are able to look them up and do quality research. However, they often end up failing ARE exams because they cannot memorize the common ARE knowledge needed on the day of the exam. ARE exams are NOT open-book exams, and you cannot look up information during the exam.

The **process of memorization** is like **filling a cup with a hole at the bottom**: You need to fill it faster than the water leaks out at the bottom, and you need to constantly fill it; otherwise, it will quickly be empty.

Once you memorize something, your brain has already started the process of forgetting it. It is natural. That is how we have enough space left in our brain to remember the really important things.

It is tough to fight against your brain's natural tendency to forget things. Acknowledging this truth and the fact that you cannot memorize everything you read, you need to focus your limited time, energy, and brainpower on the most important issues.

The biggest danger for most people is that they memorize the information in the early stages of their exam preparation but forget it before or on the day of the exam and still THINK they remember.

Most people fail the exam NOT because they cannot answer the few "advanced" questions on the exam, but because they have read the information and can NOT recall it on the day of the exam. They spend too much time preparing for the exam, drag the preparation process on for too long, seek too much information, go to too many websites, do too many practice questions and too many mock exams (one or two sets of mock exams can be good for you), and **spread themselves too thin**. They end up **missing the most important information** of the exam, and they fail.

The ARE Mock Exam series along with the tips and methodology in each of the books will help you find and improve your weakness areas, MEMORIZE the most important aspects of the test to pass the exam ON THE FIRST TRY.

So, if you have a lot of time to prepare for the ARE exams, you should plan your effort accordingly. You want your ARE knowledge to peak at the time of the exam, not before or after.

For example, if you have two months to prepare for a very difficult ARE exam, you may want to spend the first month focused on reading and understanding all of the study materials you can find as your **initial** review. During this first month, you can start memorizing after you understand the materials as long as you know you HAVE to review the materials again later to retain them. If you have memorized something once, it is easier to memorize it again later.

Next, you can spend two weeks focused on memorizing the material. You need to review the material at least three times. You can then spend one week on mock exams. The last week before the exam, focus on retaining your knowledge and reinforcing your weakest areas. Read the mistakes that you have made and think about how to avoid them during the real exam. Set aside a mock exam that you have not taken and take it seven days before test day. This will alert you to your weaknesses and provide direction for the remainder of your studies.

If you have one week to prepare for the exam, you can spend two days reading and understanding the study material, two days repeating and memorizing the material, two days on mock exams, and one day retaining the knowledge and enforcing your weakest areas.

The last one to two weeks before an exam is absolutely critical. You need to have the "do or die" mentality and be ready to study hard to pass the exam on your first try. That is how some people are able to pass an ARE exam with only one week of preparation.

17. Timing of review: the 3016 rule; memorization methods, tips, suggestions, and mnemonics

Another important strategy is to review the material in a timely manner. Some people say that the best time to review material is between 30 minutes and 16 hours (the **3016** rule) after you read it for the first time. So, if you review the material right after you read it for the first time, the review may not be helpful.

I have personally found this method extremely beneficial. The best way for me to memorize study materials is to review what I learned during the day again in the evening. This, of course, happens to fall within the timing range mentioned above.

Now that you know the **3016** rule, you may want to schedule your review accordingly. For example, you may want to read new study materials in the morning and afternoon, then after dinner do an initial review of what you learned during the day.

OR

If you are working full time, you can read new study materials in the evening or at night and then get up early the next morning to spend one or two hours on an initial review of what you learned the night before.

The initial review and memorization will make your final review and memorization much easier.

Mnemonics is a very good way for you to memorize facts and data that are otherwise very hard to memorize. It is often arbitrary, or illogical but it works.

A good mnemonic can help you remember something for a long time or even a lifetime after reading it just once. Without mnemonics, you may read the same thing many times and still not be able to memorize it.

There are a few common types of mnemonics:

1) **Visual** mnemonics link what you want to memorize to a visual image.
2) **Spatial** mnemonics link what you want to memorize to a space, and the order of things in it.
3) **Group** mnemonics break up a difficult piece into several smaller and more manageable groups or sets, in which to memorize. One example is the grouping of the 10-digit phone number into three groups in the United States. This makes the number much easier to memorize.
4) **Architectural** mnemonics are a combination of visual mnemonics, spatial mnemonics and group mnemonics.

Imagine you are walking through a building several times, along the same path. You should be able to remember the order of each room. You can then break up the information that you want to remember and link this to several images, and then imagine you hang these images on walls of various rooms. You should be able to easily recall each group in an orderly manner by imagining you are walking through the building again on the same path and looking at the images hanging on walls of each room. When you look at the images on the wall, you can easily recall the related information.

You can use your home, office, or another building that you are familiar with to build an architectural mnemonic to help you to organize the things you need to memorize.

5) **Association** mnemonics allow you to associate what you want to memorize with a sentence, a similarly pronounced word, or a place you are familiar with, etc.
6) **Emotion** mnemonics use emotion to fix an image in your memory.
7) **First letter** mnemonics use the first letter of what you want to memorize to construct a sentence or acronym. For example, "**Roy G. Biv**" can be used to memorize the order of the 7 colors of the rainbow, it is composed of the first letter of each primary color.

You can use **association** mnemonics and memorize them as all the plumbing fixtures for a typical home, PLUS Urinal.

OR

You can use "Water S K U L" (**first letter** mnemonics) to memorize them:

Water Closets
Shower
Kitchen Sinks
Urinal
Lavatory

18. The importance of good and effective study methods

There is a saying: Give a man a fish, feed him for a day. Teach a man to fish, feed him for a lifetime. I think there is some truth to this. Similarly, it is better to teach someone HOW to study than just give him good study materials. In this book, I give you good study materials to save you time, but more importantly, I want to teach you effective study methods so that you can not only study and pass ARE exams, but also so that you will benefit throughout the rest of your life for anything else you need to study or achieve. For example, I give you samples of mnemonics, but I also teach you the more important thing: HOW to make mnemonics.

Often in the same class, all the students study almost the SAME materials, but there are some students that always manage to stay at the top of the class and get good grades on exams. Why? One very important factor is that they have good study methods.

Hard work is important, but it needs to be combined with effective study methods. I think people need to work hard AND work SMART to be successful at their work, career, or anything else they are pursuing.

19. The importance of repetition: read this book at least three times

Repetition is one of the most important tips for learning. That is why I have listed it under a separate title. For example, you should treat this book as part of the core study materials for the ARE exams and you need to read this book at least three times to get all of its benefits:

1) The first time you read it, it is new information. You should focus on understanding and digesting the materials, and also do an initial review with the **3016** rule.
2) The second time you read it, focus on reading the parts I have already highlighted AND you have highlighted (the important parts and the weakest parts for you).
3) The third time, focus on memorizing the information.

Remember the analogy of the memorization process as **filling a cup with a hole on the bottom**?
Do NOT stop reading this book until you pass the real exam.

20. The importance of a routine

A routine is very important for studying. You should try to set up a routine that works for you. First, look at how much time you have to prepare for the exam, and then adjust your current routine to include exam preparation. Once you set up the routine, stick with it.

For example, you can spend from 8:00 a.m. to 12:00 noon, and 1:00 p.m. to 5:00 p.m. on studying new materials, and 7:00 p.m. to 10:00 p.m. to do an initial review of what you learned during the daytime. Then, switch your study content to mock exams, memorization and retention when it gets close to the exam date. This way, you have eleven hours of exam preparation every day. You can probably pass an ARE exam in one week with this method. Just keep repeating it as a way to retain your architectural knowledge.

OR
You can spend 7:00 p.m. to 10:00 p.m. on studying new materials, and 6:00 a.m. to 7:00 a.m. to do an initial review of what you learned the evening before. This way, you have four hours for exam preparation every day. You can probably pass an ARE exam in two weeks with this preparation schedule.

A routine can help you to memorize important information because it makes it easier for you to concentrate and work with your body clock.

Do NOT become panicked and change your routine as the exam date gets closer. It will not help to change your routine and pull all-nighters right before the exam. In fact, if you pull an all-nighter the night before the exam, you may do much worse than you would have done if you kept your routine.

All-nighters or staying up late are not effective. For example, if you break your routine and stay up one-hour late, you will feel tired the next day. You may even have to sleep a few more hours the next day, adversely affecting your study regimen.

21. The importance of short, frequent breaks and physical exercise
Short, frequent breaks and physical exercise are VERY important for you, especially when you are spending a lot of time studying. They help relax your body and mind, making it much easier for you to concentrate when you study. They make you more efficient.

Take a five-minute break, such as a walk, at least once every two hours. Do at least thirty minutes of physical exercise every day.

If you feel tired and cannot concentrate, stop, go outside, and take a five-minute walk. You will feel much better when you come back.

You need your body and brain to work well to be effective with your studying. Take good care of them. You need them to be well-maintained and in excellent condition. You need to be able to count on them when you need them.

If you do not feel like studying, maybe you can start a little bit on your studies. Just casually read a few pages. Very soon, your body and mind will warm up and you will get into study mode.

Find a room where you will NOT be disturbed when you study. A good study environment is essential for concentration.

22. **A strong vision and a clear goal**
    You need to have a strong vision and a clear goal to master the architectural knowledge needed to become an architect in the shortest time. This is your number one priority. You need to master the architectural knowledge BEFORE you do sample questions or mock exams, except for "truth-finding" exams or exercises at the very beginning of your exam prep. This will make the process much easier. Everything we discuss is to help you achieve this goal.

    As I have mentioned on many occasions, and I say it one more time here because it is so important:

    The amount of architectural knowledge and information you can understand, digest, memorize, and firmly retain is what matters, not how many books you read or how many sample tests you have taken. The books and sample tests will NOT help you if you cannot understand, digest, memorize, and retain the important information for the ARE exam.

    Cherish your limited time and effort and focus on the most important information.

23. **Codes and standards used in this book**
    We use the following codes and standards:
    American Institute of Architects, Contract Documents, Washington, DC; ADA Standards for Accessible Design, ADA; Various International Codes by ICC. See Appendixes for more information.

24. **Where can I find study materials on architectural history?**
    Every ARE exam may have a few questions related to architectural history. The following are some helpful links to FREE study materials on the topic:
    http://issuu.com/motimar/docs/history_synopsis?viewMode=magazine

25. **Which sections of *Architectural Handbook of Professional Practice* (*AHPP*) should I read?**
    Architectural Handbook of Professional Practice (*AHPP*) is a very good resource for architectural practice, but it has over 1,000 pages, and can be overwhelming for people studying for the Architectural Registration Exams (ARE). To solve this problem, the publisher of *AHPP*, John Wiley & Sons, Inc., enlisted NarmourWright Design to set up a navigation document to help candidates studying for the ARE exams. Candidates for these exams can now just focus on reading the sections of the *AHPP* related to the division of the ARE exam they are taking. Here is a link to the navigation documents for each division of the ARE exam: http://narmourwright.com/wp/index.php/wiley-publishing

Chapter Two

Overview of Programming & Analysis (PA) Division

A. General Information

1. Exam content

The PA division of the ARE has 95 questions which cover the following different areas.

Sections	Target Percentage	Expected Number of Items
Section 1: Environmental & Contextual Conditions	14-21%	13-20
Section 2: Codes & Regulations	16-22%	15-21
Section 3: Site Analysis & Programming	21-27%	20-26
Section 4: Building Analysis & Programming	37-43%	35-41

Figure 2.1 Exam format & time

Note:
After December 13, 2020, the number of questions will be reduced. See Figure 1.3.

The exam content can be further broken down as follows:
Section 1: Environmental & Contextual Conditions (14-21%)

- Evaluate site-specific environmental and socio-cultural opportunities (A/E)
- Evaluate site-specific environmental constraints (A/E)
- Determine optimal use of onsite resources by incorporating sustainability principles (U/A)

Section 2: Codes & Regulations (16-22%)

- Identify relevant code requirements for building and site types (U/A)
- Identify relevant zoning and land use requirements (U/A)
- Identify relevant local and site-specific requirements (U/A)

Section 3: Site Analysis & Programming (21-27%)

- Evaluate relevant qualitative and quantitative attributes of a site as they relate to a program (A/E)
- Synthesize site reports with other documentation and analysis (A/E)
- Analyze graphical representations regarding site analysis and site programming (A/E)

Section 4: Building Analysis & Programming (37-43%)

- Evaluate relevant qualitative and quantitative attributes of a new or existing building as they relate to the program (A/E)
- Evaluate documentation, reports, assessments, and analyses to inform the building program (A/E)
- Identify and prioritize components of the building program (A/E)
- Assess spatial and functional relationships for the building program (A/E)
- Recommend a preliminary project budget and schedule (U/A)
- Identify alternatives for building and structural systems for given programmatic requirements, preliminary budget, and schedule (U/A)
- Analyze graphical representations regarding building analysis and building programming (A/E)

2. **Official exam guide and reference index for the Programming & Analysis (PA) division**

NCARB published the exam guides for all ARE 5.0 division together as *ARE 5.0 Handbook*.

You need to read the official exam guide for this division at least three times and become very familiar with it. The real exam is VERY similar to the sample questions in the handbook.

You can download the official *ARE 5.0 Handbook* at the following link:
https://www.ncarb.org/sites/default/files/ARE5-Handbook.pdf

Note:
We suggest you study the official ARE 5.0 Handbook first, and then other study materials, and then come back to Handbook again several days before the real ARE exam.

B. The Most Important Documents/Publications for Programming & Analysis (PA) Division of the ARE Exam

1. **Official NCARB list of references for the Programming & Analysis (PA) division with our comments and suggestions**

You can find the NCARB list of references for this division in the Appendixes of this book and the *ARE 5.0 Handbook.*

Note:
*While many of the MC questions in the real ARE exam **focus on design concepts**, there are **some questions requiring calculations**.*

In the ARE exams, it may be a good idea to skip any calculation question that requires over 2 minutes of your time; just pick a guess answer, mark it, and come back to calculate it at the end. This way, you have more time to read and answer other easier questions correctly.

A calculation question that takes 20 minutes to answer will gain the same number of points as a simple question that ONLY takes 2 minutes.

If you spend 20 minutes on a calculation question earlier, you risk losing the time to read and answer ten other easier questions, which could result in a loss of ten points instead of one.

The following is the NCARB list of top references for this division. For a longer list of relevant reference materials, please see the reference matrix at the end of this book.

Publications

Architectural Graphic Standards
The American Institute of Architects
John Wiley & Sons, 11th edition (2007) and 12th edition (2016)

Problem Seeking: An Architectural Programming Primer
William M. Peña and Steven A. Parshall
John Wiley & Sons, latest edition

Site Planning and Design Handbook
Thomas H. Russ
McGraw-Hill, 2nd edition, 2009

Codes

ADA Standards for Accessible Design
U.S. Department of Justice, 2010

ICC A117.1-2009 Accessible and Usable Buildings and Facilities
International Code Council, 2010

International Building Code (2018)
International Code Council, 2017

Note:

- *The latest version of IBC is the 2018 version. For your convenience, we provide a link to the free online version of the IBC. The basic IBC content stays the same for many years. If the codes are updated by NCARB in the future, you just need to go to the root directory and find the latest version of the codes:*
  http://codes.iccsafe.org

- *You need to spend a large percentage (at least 20%) of your prep time on IBC and become familiar with it. You do not need to force yourself to memorize the numbers and all the detail. Just reading it a few times and becoming familiar with the information should be adequate.*

AIA Contract Documents
None of the standard list of AIA Contract Documents related to the ARE have specific content covered in the Programming & Analysis division.

The following are some extra study materials that are useful if you have some additional time and want to learn more. If you are tight on time, you can simply look through them and focus on the sections that cover your weaknesses:

2. **Construction Specifications Institute (CSI) MasterFormat & *Building Construction***
Become familiar with the new 6-digit CSI Construction Specifications Institute (CSI) MasterFormat as there may be a few questions based on this publication. Make sure you know which items/materials belong to which CSI MasterFormat specification section, and memorize the major section names and related numbers. For example, Division 9 is Finishes, and Division 5 is Metal, etc. Another one of my books, *Building Construction*, has detailed discussions on CSI MasterFormat specification sections.

Mnemonics for the 2004 CSI MasterFormat

The following is a good mnemonic, which relates to the 2004 CSI MasterFormat division names. Bold font signals the gaps in the numbering sequence.

This tool can save you lots of time: if you can remember the four sentences below, you can easily memorize the order of the 2004 CSI MasterFormat divisions. The number sequencing is a bit more difficult, but can be mastered if you remember the five bold words and numbers that are not sequential. Memorizing this material will not only help you in several divisions of the ARE, but also in real architectural practice

Mnemonics (pay attention to the underlined letters):
Good students can memorize material when teachers order.
F students earn F's simply 'cause **forgetting** principles have **an** effect. (21 and 25)
C students **end** everyday understanding things without memorizing. (31)
Please make professional pollution prevention inventions **everyday**. (40 and 48)

1-Good.................................. General Requirements
2-Students............................. (Site) now Existing Conditions
3-Can..................................... Concrete
4-Memorize............................ Masonry
5-Material .... Metals
6-When................................... Woods and Plastics
7-Teachers.............................. Thermal and Moisture
8-Order................................... Openings

9-F.. Finishes
10-Students............................ Specialties
11-Earn.................................. Equipment

12-F's......................................Furnishings

13-Simply...............................Special Construction

14-'Cause................................Conveying

21-Forgetting .... Fire

22-Principles.......................... Plumbing

23-Have................................ HVAC

25-An....................................Automation

26-Effect................................ Electric

27-C...................................... Communication

28-Students............................ Safety & Security

31-End..................................Earthwork

32-Everyday...........................Exterior

33-Understanding ....Utilities

34-Things...............................Transportation

35-Without Memorizing........ Waterways and Marine

40-Please................................Process Integration

41-Make.................................Material Processing and Handling Equipment

42-Professional...................... Process Heating, Cooling, and Drying Equipment

43-Pollution........................... Process Gas and Liquid Handling, Purification and Storage Equipment

44-Prevention.........................Pollution Control Equipment

45-Inventions......................... Industry-Specific Manufacturing Equipment

48-Everyday..........................Electrical Power Generation

Note:

There are 49 CSI divisions. The "missing" divisions are those “reserved for future expansion” by CSI. They are intentionally omitted from the list.

Chapter Three

Environmental & Contextual Conditions (14-21%)

NCARB Requirements
In this section, you'll evaluate a project site and identify both the opportunities and constraints that may impact future development.

According to NCARB, Environmental & Contextual Conditions includes the following sections:

A. Evaluate site-specific environmental and socio-cultural opportunities (A/E)

NCARB Requirements
You will need to be able to analyze a project site and identify opportunities that could be incorporated into future site and building development. These opportunities may include alternative energy resources, natural landscape features, microclimates, and neighborhood context, along with other environmental, social, and cultural conditions.

Note:
A/E and U/A are the levels of your knowledge or understanding of these sections expected by NCARB.

***A/E** means **Analyze/Evaluate**. It signifies an objective requiring the performance of a concept(s) using non-standard or non-predictive mental processing.*

***U/A** means **Understand/Apply**. It signifies an objective requiring the performance of a concept(s) using standard mental processing.*

1. Wind

Prevailing wind is an important factor in architectural design. Buildings can be oriented to utilize the summer breeze and avoid the cold winter wind. Prevailing wind changes with locations and seasons. A good architect should also consider the **prevailing wind** of the areas that he works in. He should incorporate this into his overall design concept. For example, in most Asian countries in the Northern Hemisphere, the prevailing wind comes from the southeast in the summer, while the prevailing summer wind is from southwest on the West Coast of America. The overall planning layout shall not block the dominant summer wind of a site to the buildings.

The **troposphere** is the lowest layer of Earth's atmosphere, and is also where nearly all weather conditions take place. In general, wind speed increases with height from the ground surface to the upper troposphere. This is for two reasons:

- A higher slant creates a greater pressure gradient between the warm and cold air and thus creates stronger wind.
- Lower wind speed, especially near the ground, is due to surface friction.

Microclimates can also generate wind. The wind speed on the upper portion of a hill can be 20% higher than that on the flat ground. The windward side of a hill has stronger wind than the leeward side. At night, cold air flows down the slope of a hill and forms a cold pocket at the foothill, called **inversion**, which holds cold air in place with warm air above it, trapping pollution and dust.

Large bodies of water, like oceans, lakes, rivers, or reservoirs can modify the microclimate and generate wind. During the day, warm air rises over the warm land and causes a breeze from the water. At night, the pattern is reversed.

Buildings and trees can also modify wind patterns. Rows of trees planted at column widths of 50 ft. to 150 ft. can form a wind break and reduce the wind speed from 30% to 60% for a distance of up to ten times the tree height. To use a windbreak effectively, an architect needs to understand what landscaping strategies will work best in that regional climate and microclimate.

In the Northern Hemisphere, if wind is the only factor of consideration, the best location to place a building is on a south, southeast, or southwest facing slope. The building should be in the middle or near the top of the slope, instead of at the bottom or top of the slope.

2. **Sun**

Earth's **tilted axis** causes the seasons. When the North Pole tilts toward the Sun, it is summer in the Northern Hemisphere, and winter in the Southern Hemisphere. And when the South Pole tilts toward the Sun, it is winter in the Northern Hemisphere, and summer in the Southern Hemisphere.

Solar altitude is the angle of the sun relative to the Earth's horizon. The altitude is zero at sunrise and sunset and can reach a maximum of 90 degrees (directly overhead) at noon in latitudes near the equator.

The **solar azimuth angle** is the compass direction from which sunlight is coming. There are different definitions. For solar energy applications, the most commonly accepted convention is that the angle is measured clockwise from due north, so east is 90°, south is 180°, and west is 270°. At solar noon, the sun is always directly south in the Northern Hemisphere and directly north in the Southern Hemisphere.

In summer, the sun is higher in the sky and sunlight travels a shorter distance through the atmosphere which normally absorbs and reflects a portion of the energy. Light also concentrates in a smaller area during summer (compare a flashlight beam hitting a wall at an angle versus the same beam hitting a wall perpendicularly), and days are longer. That is why summer is warmer than winter.

The following information is from the National Weather Service:

1) **The Equinox (Vernal & Autumnal)**

There are only two times of the year when the Earth's axis is tilted neither toward nor away from the sun, resulting in a "nearly" equal amount of daylight and darkness at all latitudes. These events are referred to as Equinoxes. The word equinox is derived from two Latin words - *aequus* (equal) and *nox* (night). At the equator, the sun is directly overhead at noon on these two equinoxes. The "nearly" equal hours of day and night is due to refraction of sunlight or a bending of the light's rays that causes the sun to appear above the horizon when the actual position of the sun is below the horizon. Additionally, the days become a little longer at the higher latitudes (those at a distance from the equator) because it takes the sun longer to rise and set. Therefore, on the equinox and for several days before and after the equinox, the length of day will range from about 12 hours and six and one-half minutes at the equator, to 12 hours and 8 minutes at 30 degrees latitude, to 12 hours and 16 minutes at 60 degrees latitude.

2) **The Solstices (Summer & Winter)**

The summer solstice occurs at the moment the earth's tilt toward from the sun is at a maximum. Therefore, on the day of the summer solstice, the sun appears at its highest elevation with a noontime position that changes very little for several days before and after the summer solstice. The summer solstice occurs when the sun is directly over the Tropic of Cancer, which is located at 23.5° latitude North, and runs through Mexico, the Bahamas, Egypt, Saudi Arabia, India, and southern China. For every place north of the Tropic of Cancer, the sun is at its highest point in the sky and this is the longest day of the year.

The winter solstice marks the shortest day and longest night of the year. In the Northern Hemisphere, it occurs when the sun is directly over the Tropic of Capricorn, which is located at 23.5° south of the equator and runs through Australia, Chile, southern Brazil, and northern South Africa.

3) **The Seasons**

We all know that the Earth makes a complete revolution around the sun once every 365 days, following an orbit that is elliptical in shape. This means that the distance between the Earth and Sun, which is 93 million miles on average, varies throughout the year. During the first week in January, the Earth is about 1.6 million miles closer to the sun. This is referred to as the **perihelion**. The **aphelion**, or the point at which the Earth is about 1.6 million miles farther away from the sun, occurs during the first week in July. This fact may sound contradictory to what we know about seasons in the Northern Hemisphere, but actually the difference is not significant in terms of climate and is NOT the reason why we have seasons. **Seasons are caused by the fact that the Earth is tilted on its axis by 23.5°.** The tilt's orientation with respect to space does not change during the year; thus, the Northern Hemisphere is tilted toward the sun in June and away from the sun in December, as illustrated in figure 3.1.

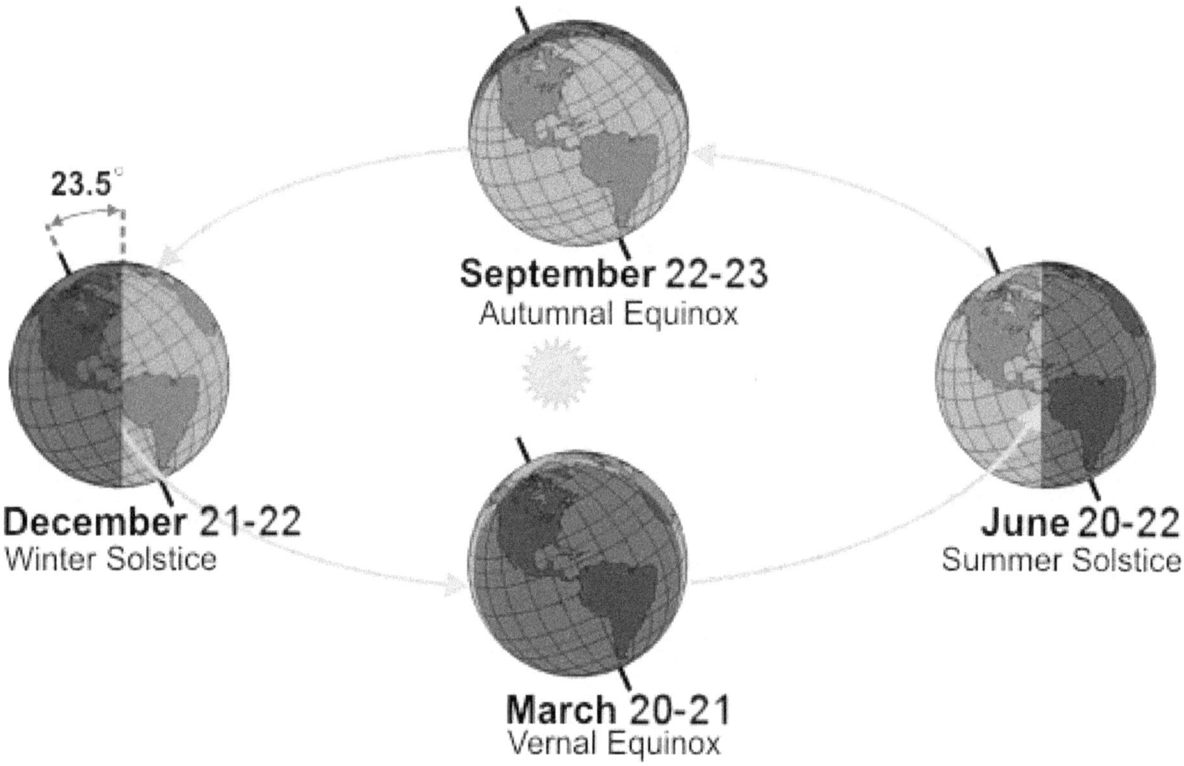

Figure 3.1 Earth orbit

4) **The Relationship Between Length of Day and Temperature**
There is a lag between the longest day of the year and the warmest average temperatures for most mid and high latitude locations.

In northern OH/northwest PA, the maximum daily temperature occurs nearly 3 weeks later in mid-July. Just as the warmest part of the day usually occurs several hours after noon, when the sun is highest in the sky, so too does the warmest part of the summer lags the summer solstice. This lag is due to the time required for ground and water to heat up. Average temperatures continue to climb until the sun drops lower in the sky. While the effect is evident in a daily temperature plot, it is more readily apparent by looking at changes in the monthly average temperature. In Cleveland, July averages 3.3°F degrees higher than June, with August also warmer than June by 2.8°F, even though the length of days in August is considerably less than the length of days in June.

See the National Weather Service's website at the following link for more information:
https://www.weather.gov/cle/Seasons#

In the Northern Hemisphere, south-facing slopes tend to be warmer, especially in winter, compared to slopes facing other directions.

Surface material affects the amount of solar radiation absorbed or reflected. **Albedo** is the proportion of radiation that is reflected by a surface, and ranges from 0 to 1.0. The albedo of a black surface that does not reflect anything is 0. The albedo of a mirror that reflects everything is 1.0. Grass and vegetation have low albedo. Concrete pavement and snow have high albedo.

Albedo is closely related to reflectivity. **Reflectivity** is the property of reflecting radiation or light, especially reflectance as measured independently from the thickness of a material, and ranges from 0 to 1.0. Albedo is a measure of *energy* and reflectivity is a *property* of a material.

The **emissivity** of the surface of a material measures its effectiveness in emitting *energy* as thermal radiation, and ranges from 0 to 1.0.

For an *opaque* object, **emissivity + albedo = 1.0**

The **solar reflectance index (SRI)** is a measure of a surface's *ability* to reflect solar heat, as shown by a small temperature rise. The SRI of a standard black surface (reflectance 0.05, emittance 0.90) is 0, and the SRI of a standard white surface (reflectance 0.80, emittance 0.90) is 100. The SRI of a perfect reflected surface is 122.

Solar reflectance is the *ability* of a material to reflect solar energy from its surface back into the atmosphere. The **SR value** is a number from 0 to 1.0. A value of 0 indicates that the material absorbs all solar energy and a value of 1.0 indicates total reflectance.

The **orientation of a building** has a profound impact on its energy gains and losses. For instance, at 40° latitude in the Northern Hemisphere during winter, a southern exposure receives three times the solar energy as the west or east exposure. During summer, the total energy received by the west and east sides is about twice as much as the total of the southern and northern exposures. In the Northern Hemisphere, it is best to orient a building's main elevation towards the south or slightly west or east of south.

The USGBC uses the LEED system to rate green buildings. The proper use of SRI values on the roof and the SR values of non-roof materials can reduce the **heat island** effect and may contribute points to the LEED rating system.

Note:
*An urban **heat island is** an urban area with a significantly higher temperature than the surrounding rural areas due to human activities.*

3. Plants and buildings

Plants can be used to provide shade for buildings so as to reduce the temperature of building walls and roofs and to save energy and air-conditioning cost in the summertime.

1) **To the east of buildings**
To the east of the buildings, tall tree with dense canopies can be planted to block sunlight coming from a high angle, while not blocking the view of the sunrise. In the early morning hours, when the sun is rising at a low angle, the sunlight is still fairly gentle, therefore, there is no need to block it. In late morning hours, the sun is at a high angle and the sunlight is very strong and produces a great amount of heat; tall trees with dense canopies can block the strong sunlight at a high angle (figure 3.2, figure 3.3).

2) **To the south of buildings**
In the Northern Hemisphere, we probably would plant tall and high crown deciduous trees to the south of buildings. This is because the sunlight is at a high angle in the summer, while at a low angle in the winter. Deciduous trees have full foliage and can block high-angle sunlight to reduce heat in the buildings in the summer. They have no leaves in the winter and can allow sunlight to penetrate through the branches and south-facing windows and enter rooms to raise room temperature at a low angle.

3) **To the west of buildings**
Westerly sunlight is one of the major reasons that buildings overheat. To the west of buildings, plants should be able to block the afternoon sunlight. One commonly used technique is to use a trellis with climbing plants to provide shade for the buildings and also create a resting area under the trellis in the summertime. Climbing plants are also used to climb up vertical trellises or fences, which are very effective in blocking low-angle late afternoon sunlight.

4) **To the north of buildings**
A chilly north wind in the winter can significantly increase the heating energy cost. Evergreen plants with dense foliage, such as cypress trees, are often planted closely to the north of buildings to form windbreaks.

5) **Rooftop gardens**
One of the Seven Wonders of the World, the Hanging Gardens of Babylon, may be one of the earliest rooftop gardens. Waterproofing is the biggest challenge for rooftop gardens. To prevent damage to the roof and the waterproofing system, plants used in rooftop gardens shall have shallow roots and modest size, and are often placed in planting pots or planters. Climbing plants and trellises are also excellent choices for rooftop gardens. One of the recent examples for rooftop gardens is located at the pool area for The Venetian Las Vegas hotel and Casino. It is a rooftop garden with a health and fitness center and a wedding chapel.

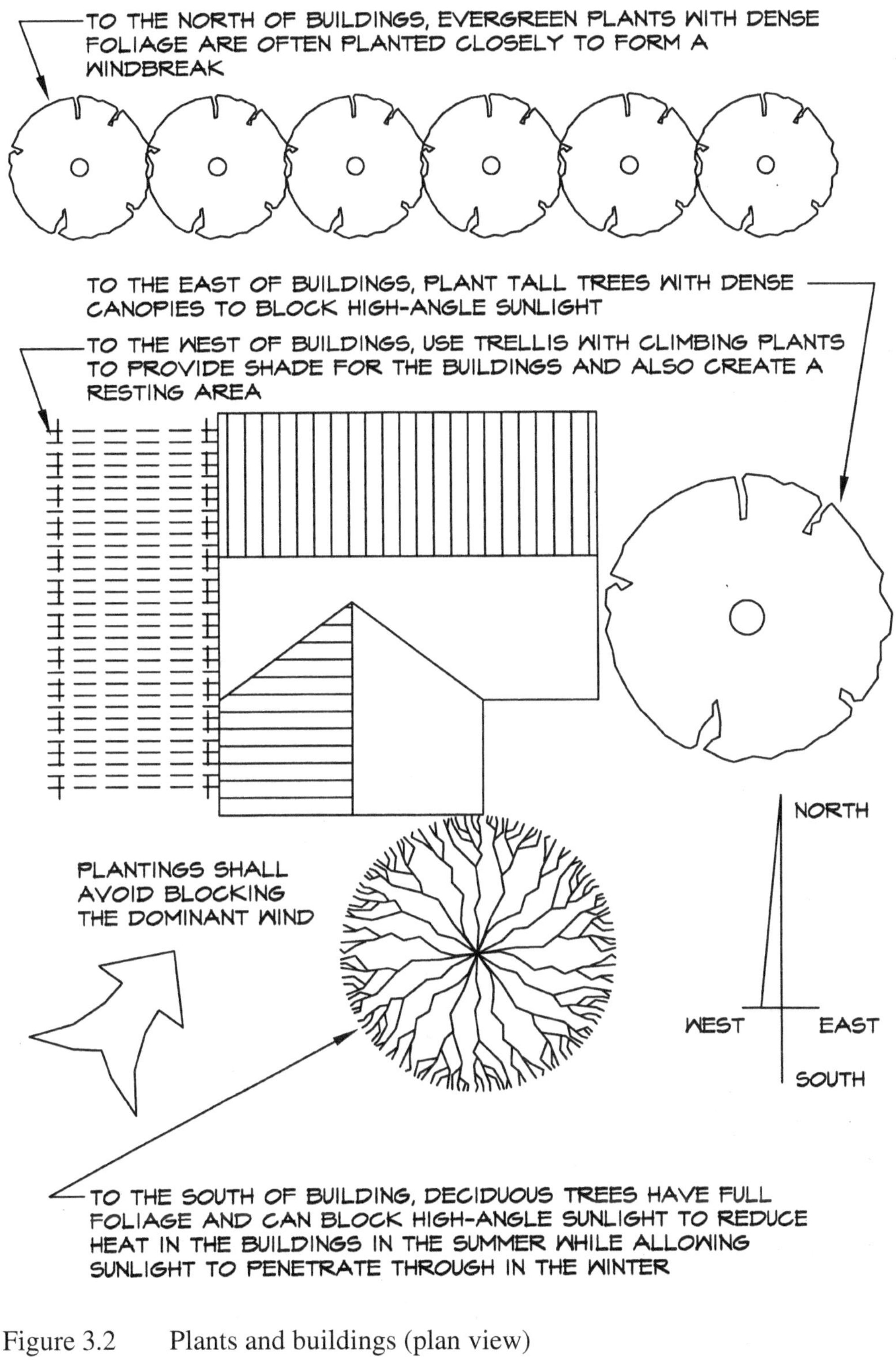

Figure 3.2 Plants and buildings (plan view)

Figure 3.3 Plants and buildings (section views)

If you are interested in more information on landscape design and vegetation, you can refer to my other book, *Planting Design Illustrated*, which is available at:
http://www.GreenExamEducation.com

4. **Shading**

In addition to plants, architectural elements such as awnings, overhangs, and trellises can provide shading too. Some shading devices also act as **reflectors**, called **light shelves**, which bounce natural light deep into a building's interiors.

Well-designed sun control and shading devices can:

- dramatically reduce building peak heat gain and cooling requirements, resulting in reductions of annual cooling energy consumption by 5% to 15%
- improve the natural lighting quality of building interiors
- improve user visual comfort by controlling glare and reducing contrast ratios, which often leads to increased satisfaction and productivity
- offer the opportunity of differentiating one building facade from another and providing human scale and interest to a design

The design of shading devices is difficult to generalize about due to the range of climates and the wide variety of buildings they are used in. However, the following design recommendations can help:

- Use fixed overhangs on south-facing glass to control the direct beam of solar radiation. Use other measures, such as low-E glazing to control indirect (diffuse) radiation.
- Minimize the amount of east and west facing glass since it is harder to shade than the south. Use vertical louvers at east and west windows.
- There is no need to shade north-facing glass in the continental United States since it receives very little direct solar gain, but this is not true in the tropics of Central and South America. Here the roof should be shaded, even if there are no skylights, because this surface is a major source of transmitted solar heat gain into the building.
- Shading effects daylighting; consider both simultaneously. A light shelf, for instance can shade lower windows and bounce natural light deeply into a room through high windows.
- Interior shading devices such as vertical louvers or Venetian blinds cannot reduce cooling loads since the solar gain has already been admitted into the room. However, these interior devices can contribute to visual acuity, visual comfort, and offer glare control.
- Study sun angles. An understanding of sun angles is critical to design including determining building orientation, placing building integrated photovoltaic (BIPV) panels or solar collectors, and selecting shading devices.
- Consider the durability and maintenance cost of shading devices.
- Shading strategies that work well at one latitude, may be completely inappropriate for other latitudes.

See **Whole Building Design Guide (WBDG)** at the following link for more information: https://www.wbdg.org/resources/sun-control-and-shading-devices

5. **Views**

An architect should study the site carefully, take full advantage of good views, and avoid unsightly areas in his design. For example, a carefully placed window can frame a vista in the distance, while a trellis can help to block the views to an unsightly loading dock.

Extensive discussions on views and design can be found in my book, *Planting Design Illustrated* (**ISBN-13:** 978-0984374199):

"An example of using views in design: the mud mounds built by Sumerians and many early gardens might have inspired the high terraces in the formal gardens in France and other countries. Formal gardens and planting design were frequently designed to be looked down upon from high terraces or windows of upper floors. This design concept was also probably because many of the more complicated formal patterns were easier to see and understand from above. It was a very ancient idea that higher is better.

This concept was almost universal in both the Western and Eastern cultures. In large-scale French gardens, high viewing terraces were frequently built for visitors to perceive the formal patterns of parterres and flowerbeds and knot gardens. Tall trees were placed at outer edges or used to frame a vista, and they were never placed close to buildings so as not to block views of the formal patterns of flowerbeds, parterre, etc.

Last but not least, formal gardens and planting design are often symmetrical, formal, and balanced. After so many generations of development and trial and error, they have become mature. They are beautiful and loved by many people. Our task is to learn and inherit from the past, analyze and absorb the essence of design, and use them for today's needs…

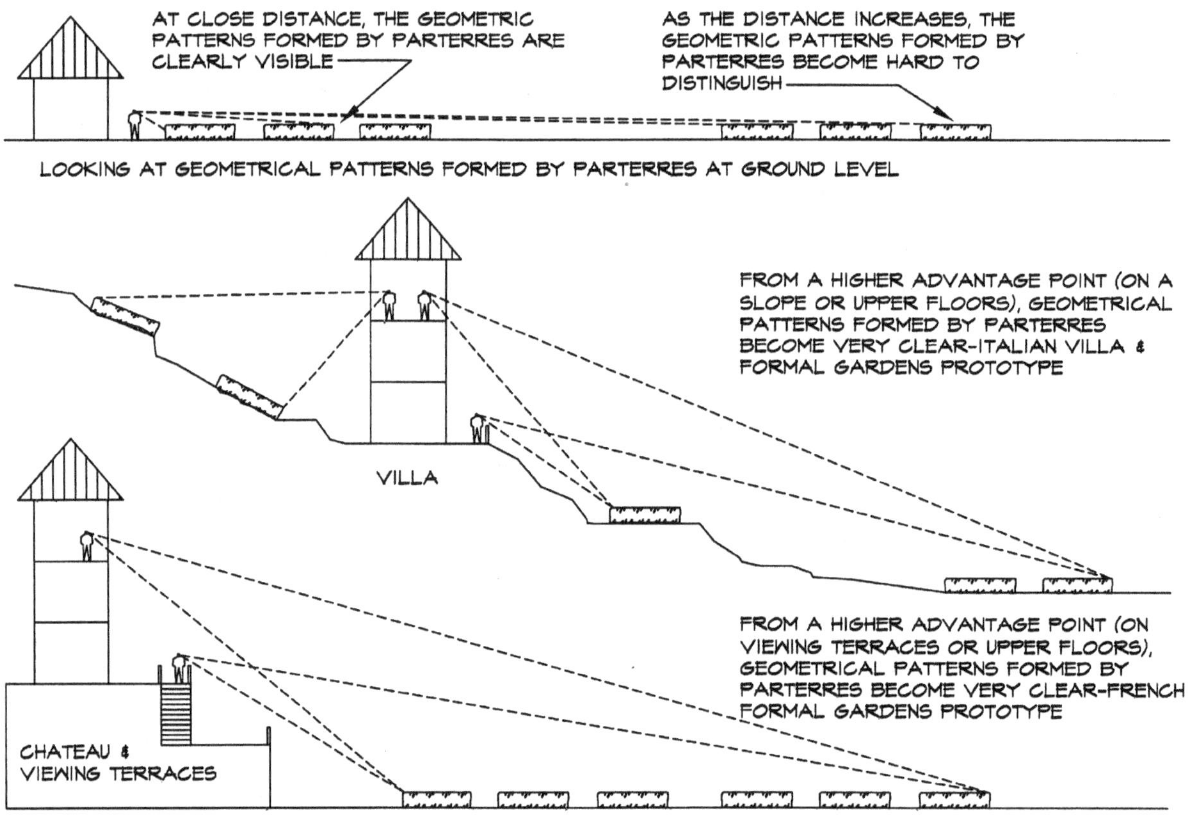

Figure 3.4 View line studies of formal gardens and architectural design

During the Renaissance, Italian humanist and architect *Leon Battista Alberti* (1404-72) wrote *De re aedificatoria* in 1452, and this book was published thirty years later. He drew extensively from the writings of Pliny the Younger, sometimes almost copying the text verbatim, especially on how to locate villas and gardens: "villas and their gardens should be located on a hillside to obtain views, sun and wind. The slope should be gentle and visitors should hardly notice the ascent until they see the view as they arrive at the villa." Like Pliny the Younger, Alberti also recommended the use of scented evergreens and topiary, fountains, flowing water, statues, grottos, and stone vases.

Many formal and villa gardens were created in Italy during the Renaissance. They had several common features and obeyed numerous basic design principles. The formal gardens were designed for viewing from the main rooms of the villa.

Since the designers probably realized that the formal gardens look best from above, they placed the villa at a higher location (typically the slope of a hill), and the villas were often built to be three or four stories high and placed on a terrace to make it even higher (figure 3.4). The gardens were placed at various lower levels of terraces or slope. Lower flowerbed or parterres were placed closer to the villa and higher plants were placed farther away from the villa.

The villa and the formal gardens were designed as one unity and shared the same main axis with pools, fountains, and obelisks; sculptures were arranged along the axis to form a series of different spaces.

The villas and gardens were basically symmetrical to the main axis but some variations were allowed. The gardens were divided into regular, formal shapes by straight and controlled paths. The paths were framed either by cypresses or a clipped edge. Pavilions were placed at the intersections of paths.

In additional to **cross axes, diagonal axes** were also used to link various points of interest and create surprises for visitors. These diagonal axes also created the radial patterns of the paths. Early villa and formal gardens followed the ancient Roman villa and formal garden prototype and the villas were placed at a very gentle slope, but in some later gardens, the slope became steeper and grand flights of steps were used.

The designers took advantage of the slopes and differences in contour and created many water works and running streams. Water jets and fountains were used extensively. Famous examples included Villa *Medici* at *Pratolini,* Villa *Medici* at *Castello,* The *Pitti* Palace at Florence, Villa *d'Este* at *Tivoli,* Villa *Lante*, Villa *Torlonia* at *Frascati*, Villa *Garzoni* at *Collodi* and the *Isola Bella* in Lake *Maggiore*...

In 1651, *André Mollet* published *Jardin de Plaisir.* He summarized the pattern for the French formal gardens in this book: from the main rooms of the house, there was an unblocked view of the parterre, and then parterres of turf and formal woods. A statue or fountain is placed at the end of almost every walk to create a vista. Grottos, aviaries, canals, and other waterworks were other elements of the formal garden. He even suggested placing painted scenes at the end of walks to create more effective vistas. Again, the entire garden was designed for viewing from the main room of the house and the formal layout was to show man's control and ability and the power and authority of the owner.[1]

Versailles at France is one of the most famous illustrations of formal gardens and planting design (figure 3.5). It was designed by *André Le Nôtre* and *Le Brun* in the seventeenth century, and was originally a hunting park for the French king and subsequently enlarged and developed into the grand and glorious setting for the monarch, Louis XIV, the Sun King.

Louis decided to use the Sun God, Apollo, as the theme of symbolism for the garden. *Le Nôtre* and *Le Brun* were discovered by the king after they designed *Vaux-le-Vicomte* for *Nicolas Fouquet*. *Le Brun* was in charge of the symbolism and sculpture and *Le Nôtre* was in charge of the garden and planting design.

Le Nôtre's design technique was to make the entire garden visible at a glance, so it must be relatively narrow so that a man's eyes do not have to turn from side to side. In order for the garden to be impressive and to show its grand scale, the garden must be long, so that a man can stand on the uppermost viewing terrace and look on and on. The main axis of the garden is also the main axis of the buildings; it ran from the center of the palace westwards to the setting sun. This made Louis the center of the design. The Apollo fountain was placed along this axis at the end of the canal.

Le Nôtre used the radial pattern as the motif of the garden, and used the radial routes at both the main axis along the canal and the secondary axes of the smaller gardens. Fountains or statues were placed at the centers of the radial patterns. This way, a fountain or statue can serve as vista for multiple paths or walks. Each smaller garden may have its unique characteristic, but it must be subordinate to the whole design concept. Three grand avenues connect the Versailles town to Place d'Armes and the palace.

At the west side of the palace, high terraces make it possible for a visitor to see the entire garden at a glance and look at the main vista framed by trees and shrubs arranged in linear patterns and following the main axis of the palace. A number of low parterres and green lawn in various formal patterns have been placed close to the high viewing terrace. They can be looked down upon from the terraces and the upper floors of nearby buildings. Visitors can actually perceive geometric patterns formed by parterres and green lawn because they can look down at these patterns.

Beyond that, paths and canals are framed by woodland. The canal was crossed by another channel linking a menagerie and Trianon. After the canals, the ground rises steeply and thicker woods leading to the horizon frame the paths. Overall, the Garden of Versailles was not built for pedestrians, but for visitors on a horse-drawn carriage.[2]

A number of patterns we'll discuss have been used in the planting design of Versailles: axis and symmetry, linear array with vista, radial pattern, motif, repetition and rhythm, etc.

---

[1] Loxton, Howard (Editor). *The Garden: A Celebration*. David Bateman Ltd., 1991. pp. 10-60.

[2] Morris, A.E.J. *History of Urban Form: Before the Industrial Revolutions*. England: Longman Scientific and Technical, Longman Group UK Limited, 1979. Reprinted 1990. pp. 174-178.

Versailles is also a good example of the balance between unity and variations. *André Le Nôtre* used the cross axes to set up the basic pattern of the garden, yet he also used the diagonal axes to break the pattern and create surprise, interest, and climax. The smaller gardens are similar in size and most are in rectangular shapes, yet the patterns inside each garden are unique.

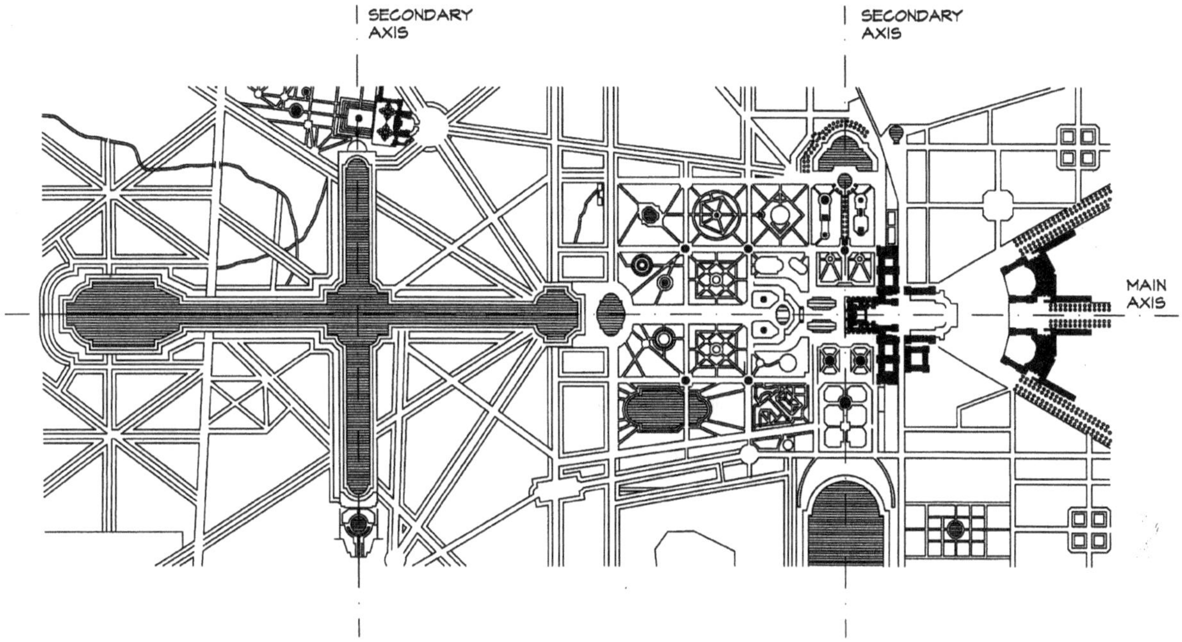

Figure 3.5 Versailles at France is one of the most famous illustrations of formal design

Versailles is like an extremely well-done symphony, with a very clear main theme, containing numerous variations. All these variations are unique and yet related and unified in some way to reinforce the main theme and create a harmonious overall planting design.

A good designer should be able to master all basic formal design patterns and use them flexibly.

Pierre L'Enfant, the planner and designer for Washington D.C., spent his childhood in Versailles. This probably had a significant impact on his concept for the overall design of Washington D.C. (figure 3.6), and explains the extensive use of diagonal axes, radial patterns, and grand avenues in the planning and design of Washington D.C...

You may consider your design project's relation to the entire city. This is the most important scale and is often missed by most designers. For example, the Mall at Washington DC is located at one of the most important axes of the city. This axis connects the Robert F. Kennedy Memorial Stadium, Lincoln Park, U.S. Capitol Hill, Washington Monument, National WWII Memorial, and Lincoln Memorial.

This design concept means it would make a lot of sense to design the planting in the Mall in a ceremonial, formal, and classical style, probably in a symmetric manner to reinforce this important axis. The size of the trees used probably need to be huge along the Mall to match the grand columns and other architectural components used in the U. S. Capitol Hill and Lincoln Memorial. The center area can be lawn or other low-rise plants so as not to block the view of U.S. Capitol Hill, Washington Monument, and Lincoln Memorial along this axis. The overall theme of the design may be grand and dignified to demonstrate the great power of the federal government (figure 3.6).

This planting design layout would also be consistent with Pierre L'Enfant's original overall city design concept in 1791 and the "Grand Avenue" (Boulevard) and wide and monumental public space that he intended to create to symbolize that America is a nation with room to grow.

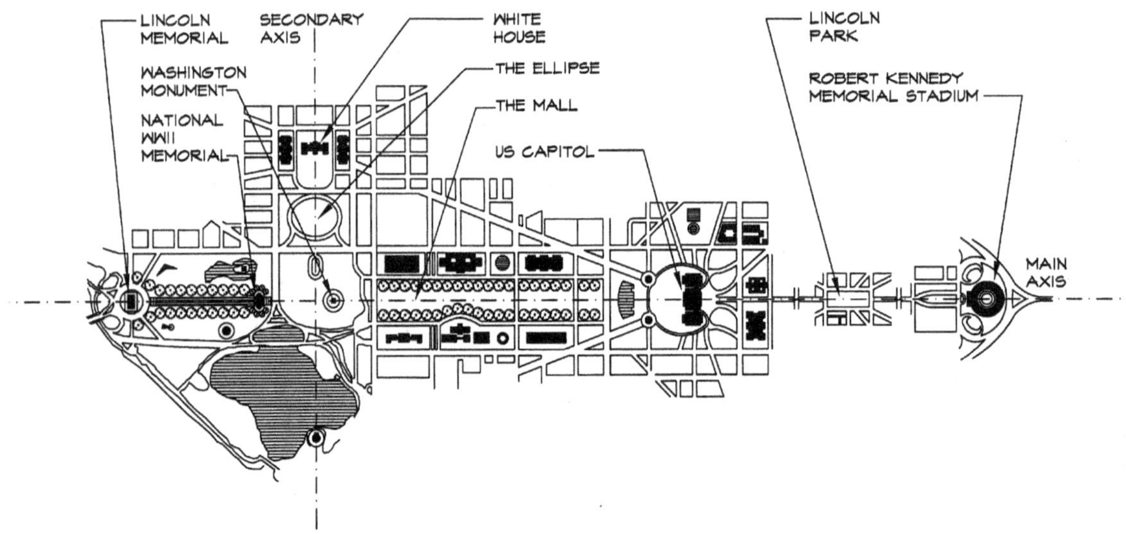

Figure 3.6 The Mall, Washington DC

6. **Water**
   Water can be a great amenity for building and site design. An architect should fully utilize any water features available at a site. Many cities in the world have started to revitalize their waterfronts and create walkable paths along these areas. Water can modify microclimate, create reflections of buildings and landscape, provide a habitat for plants and animals, and improve the environment.

   All dominant religions consider water very important and believe the element symbolizes life. They called water **"living water."** The center of many gardens are frequently water or some kind of water feature. From the pool in Egyptian courtyard gardens to the main moat in Versailles, there are numerous historical examples.

   Fallingwater by Frank Lloyd Wright is a famous example of fully utilizing the water at the site, and also using the building itself to add interest to the water. The building and water are integrated together.

7. **Neighborhood and the four design scales**
   Every design can be thought out at four different scales: at the city scale, at the plaza/neighborhood scale, at the building scale, and at the detail scale. Let's discuss each of the scales as follows:

1) **At city scale**
You may consider your planting design project's relation to the entire city. This is the most important scale and is often missed by most designers. The Mall, Washington DC we discussed previously is a good example of designing at city scale.

Central Park at New York is a global wonder of design. It has experienced many changes in the past 150 years, yet the most fundamental concept and overall layout was decided in 1858 by the original designers, Frederick Law Olmsted and Calvert Vaux.

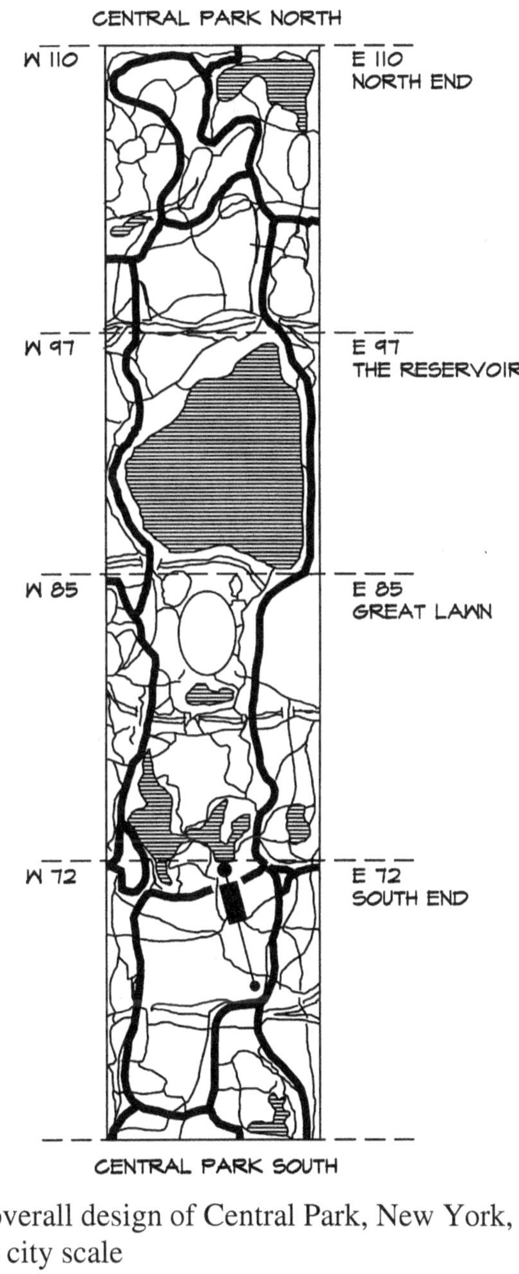

Figure 3.7 The overall design of Central Park, New York, was set to be naturalistic at the city scale

At the city scale, they decided to create a centralized great park instead of many small parks for New York City, and to make the overall design of the park be naturalistic or "picturesque" landscape to provide an opportunity for the people to escape from the stress of the city. They believed in the healing power of nature. The park has changed over time, yet the overall characteristics remain basically the same as the original design (figure 3.7).

In the early 1990s, I visited an exhibition of the design entries for the Korean Museum of Arts in Los Angeles. There were over 300 entries; many entries were from internationally known architectural firms. Every entry had excellent drawings, models, and presentations. I was wondering how the judges could pick from all these marvelous designs. It seemed that it would be a very difficult decision to pick the winner, yet all the judges came up with a unanimous vote for the winner. The winner had made their task very easy. The winning design scheme outpaced all other entries at the city scale of the design: the design concept came from Korean tradition; a huge bell covered by a classical pavilion was a center for a traditional Korean community. The designer created a huge public space/plaza for the city, covered by a huge "roof of the city." The designer created a U-shaped space enclosed by a multi-story building complex with a cover or huge "roof of the city" formed by the upper building floors. Under the cover was an auditorium in an abstract shape to simulate the huge bell covered by a classical pavilion. This designer incorporated the Korean heritage in a very modern style, and created an important public space at the city scale through his bold imagination. His creative concept at the city scale was the key to the success.

We can apply the same philosophy to our design. Architects probably need to seize every opportunity to create a meaningful public space at the city scale. When we design, we are creating an important public, semi-private, or private space. The overall layout of the design at the city scale is of paramount importance.

2) **At plaza or neighborhood scale**

Buildings or other design elements can be arranged to reinforce the design concept of the plaza or neighborhood. For example, in a plaza with a circular fountain and radial pavement patterns, the trees can be arranged in a similar and circular pattern to reinforce the plaza's overall design concept (figure 3.8). In a residential neighborhood, an architect can create a very unique effect if he can put in a little thought at the plaza or neighborhood scale.

For instance, if an architect places one tree in the front yard of each home along a residential street, this will create a tree-lined street. If all the trees are of the same or similar form, this street can be a very special street. Sometimes the street can even be named after the trees used (Maple St., Sycamore St., etc.) if it is in a new development area (figure 3.9). A new tract home development will offer designers this kind of opportunity to create some unique character at the plaza/neighborhood scale. This can usually be achieved by incorporating the planting design requirements in the **covenant, condition, and restrictions (CC and R)** of the **homeowner's associations (HOA).**

A neighborhood is a small area forming a community within a city, where people share similar needs in social activities and housing.

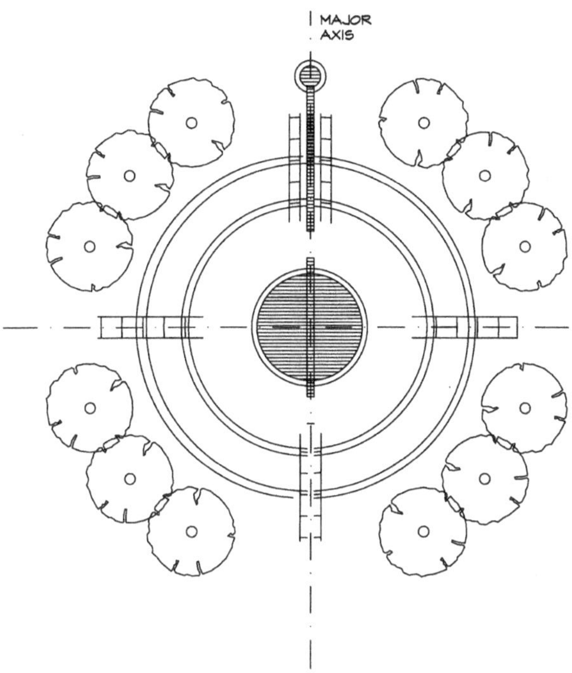

Figure 3.8 Trees at plaza scale can match the design concept of the plaza

Any new project will become an intimate part of the surroundings. An architect needs to be sensitive to the existing fabric of the neighborhood, and also the project's impact to the surroundings.

Clarence Perry developed the first neighborhood concept in 1929, mainly as a way to get people involved in the planning process and discuss common problems. His idea was to *design the elementary school as the center of the neighborhood.* The entire neighborhood was to be within walking distance of the school and major streets were placed at the boundaries of the neighborhoods so that children could walk to and from school without crossing them. This area become a basic **neighborhood unit**.

The neighborhood has become a basic unit in urban planning and design. It has become a self-sustainable unit for many basic daily functions not requiring a car and within a walking distance, and can include an elementary school, bank, grocery store, dentist, restaurant, etc. It is a manageable sized unit to plan and build a city with, and is a part of the city that people are familiar with and can identify with.

A designer should respect the existing neighborhood of a project by maintaining the existing amenities, using similar size, scale or materials, respecting views and access to important structures, and trying to fit in.

3) At building scale

You should consider other design elements such as plants' relationship to a building. The scale, proportion, and scale of plants close to a building probably need to be chosen according to the size, height, and scale of the buildings.

If a building has a large area of wall surface, it may appear to be too monotonous; plants can be used to soften the building elevation. Climbing plants (vines and other climbers) are often used to climb onto a large wall area to form a special composition, or climb along the colonnades or archway to soften a building's elevations.

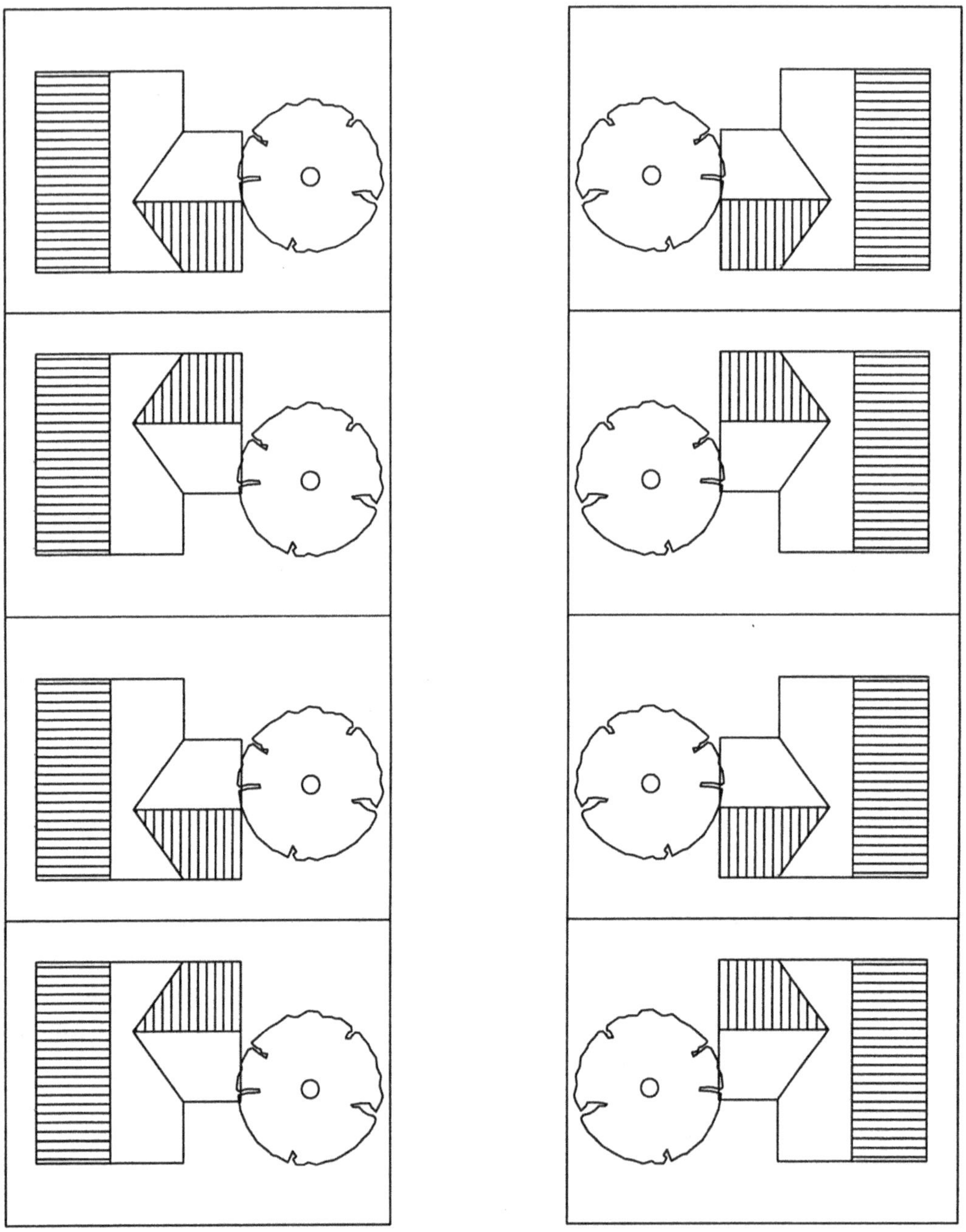

Figure 3.9 Plant one tree in front of each home to form a very unique tree-lined street

If a building's elevations appear to be too flat, high trees can be used to break the tedious skyline and create more interesting elevations. In the shade formed by canopies or other building elements, shade-loving plants can be placed.

Trees placed close to or in front of the windows of buildings (especially multi-story buildings) should have very thin foliage to avoid blocking the view line of people inside the building (figure 3.10). Trees with thick and huge tree canopies should never be placed in front of the windows of buildings. It is human nature to desire to look out of the windows of a building. Many people have complained about their view lines being blocked by trees right in front of their windows.

Figure 3.10 Plants and view lines from a building

4) At detail scale

You can utilize many properties of buildings and other design elements such as plants at detail scale. For example, you can use plants to create sound effects, take advantage of the color and aroma of plants, and emphasize the beauty of lines formed by plants' branches. It is important to remember that most of the decorative plants probably need to be placed at a close viewing distance for people to see their decorative features. For example, decorative plants may be placed right next to the walkway to achieve maximum exposure. Decorative plants may also be placed along a waterfall to form a focus of the garden. Both the decorative plants and a waterfall can be placed at a close viewing distance next to a walkway.

I once talked with a well-known designer and asked him what his secret to success was. He told me, "It is in details, God is in the details." This may be an overstatement, but it does show that details are very important to landscape and planting design.

8. Social and cultural conditions

1) Social conditions

Many architects have a strong motivation to serve the needs of the user population and public good. Social values and awareness have an impact on architectural designs too.

Social design values are sometimes in conflict with other values, and represent the continuing debates between rationalism and romanticism in architecture.

Social design values include the following categories:

a. The social change design value

This design value is a commitment to use architecture to change society for the better, and is often related to political movements and subsequent building programs, such as low-income housing programs, etc. Architects use the design as a tool to transform the built environment and its users.

b. The consultation and participation design value

This design value believes involving stakeholders in the design process is beneficial and can lead to:

- an effective use of resources for meeting social needs
- user influence in the design process and user awareness of the consequences, etc.
- up-to-date and relevant information for designers

c. The crime prevention design value

This design value believes that the architects can manipulate the built environment to reduce crime levels through the following three main strategies:

- defensible space
- crime prevention via environmental design
- situational crime prevention

d. The 'third world' design value

This is to help developing countries through architectural design (i.e. a response to the needs of the poor and destitute within the third world). This design value considers that the third world needs the development of special solutions because of its social and economic circumstances.

2) Cultural conditions

Cultural conditions, including biophilia and traditional design values, have significant impacts on architectural design.

a. Biophilic architecture

The term **"biophilia"** means "love of life or living systems." The biophilia hypothesis considers that humans have an innate tendency to seek connections with nature and other forms of life. **Philias** are the positive feelings and attractions that humans have toward organisms, species, habitats, processes, and objects in their natural surroundings.

Biophilia is refined through culture and experience, and is hypothetically the result of *biological evolution.* Our natural love for life helps sustain life.

With our technological advancements, people spend more and more time inside buildings and cars. It has become more important in conservation to reestablish a connection with nature. The solution is to provide more green spaces in and around cities and implement smart design for greener cities that integrate ecosystems into them such as biophilic cities. These green spaces can also become a part of wildlife corridors to help with territorial and migrational needs of animals.

In architecture, **biophilic design** is a sustainable design strategy reconnecting people with the natural environment, and is able to supply our inborn need of connection to the vital processes and to life.

Biophilic architecture is characterized by:

- the naturalistic dimension
- the wholeness of the site, that is, "the basic structure of the place"
- the "geometric coherency." A physical space must have a geometrical configuration able to promote the connection of human and built and natural environments.

Similarly, **biophilic space** has been defined as an environment that supports life and can strengthen psychological and sociological components. The space can:

- unburden our cognitive system, and help collect and recognize more information in the quickest and most efficient way
- avoid both the exciting and the depressive effects, and foster the optimum of our sensorial system in terms of neuro-motorial influence
- induce a strengthening in biological and emotive terms at a neural level

- support the neuro-endocryne and immunological system, especially for those people with poor health

Having a window looking out to plants or having plants in the same room as patients can help speed up the healing process of patients in hospitals.

b. Traditional design values
Traditional design values include the following categories:

a) **The tradition-based design value** is rooted in a belief that traditional designs are the preferred typology and template for buildings because they create functional and timeless designs. They can be further divided into the following categories:
- critical regionalists or traditionalists who interpret the traditional templates and typologies and who use them in an abstracted modern vocabulary
- revivalists who adhere to the most literal traditional form
- contextualists who use historical forms when the surroundings "demand" it

b) **The design value of restoration and preservation** is based on a commitment to preserve the best buildings for future generations. It tends to represent restoring a building to its initial design and is usually rooted in three perspectives:
- an archaeological perspective that preserves buildings of historical interest
- an artistic perspective that desires to preserve something of beauty
- a social perspective that desires to hold onto the familiar and reassuring

c) **The vernacular design value** is based on a belief that a simple life and its links to nature, are superior to that of modernity. The design value of vernacular includes:
- reinvigorating tradition by evoking the vernacular
- reinventing tradition by searching for new paradigms
- extending tradition by using the vernacular in a modified manner
- reinterpreting tradition by the use of contemporary idioms

B. Evaluate site-specific environmental constraints (A/E)

NCARB Requirements
You will need to analyze a project site and assess constraints and hazardous conditions that could limit a building's location and future site development. These constraints may include issues like a floodplain, unstable soil, radon, lead, a brownfield, abandoned structures, environmental policies and regulations, or vulnerabilities and risks associated with climate change. You will also need to identify strategies that could mitigate adverse site conditions.

1. Floodplain
FEMA defines a **flood** as a general and temporary condition of partial or complete inundation of normally dry land areas from:
- the overflow of inland or tidal waters;
- the unusual and rapid accumulation or runoff of surface waters from any source;

- mudslides (i.e., mudflows) which are proximately caused by flooding and are akin to a river of liquid and flowing mud on the surface of normally dry land, as when earth is carried by a current of water and deposited along the path of the current.

A flood inundates a floodplain. Most floods fall into three major categories: riverine flooding, coastal flooding, and shallow flooding. Alluvial fan flooding is another type of flooding more common in the mountainous western states.

Flood hazard areas identified on flood insurance rate maps, are identified as a **special flood hazard area (SFHA)**. SFHA is defined as the area that will be inundated by a flood event having a one percent chance of being equaled or exceeded in any given year. The one percent annual chance flood is also referred to as a base flood or 100-year flood.

The most common flood damages arise from:

- direct damage during a flood from inundation, waves, high velocity flow, erosion, sedimentation, and/or flood-borne debris
- degradation of building materials, either during or after the flood
- contamination of the building due to mold or flood-borne substances

These damages can be minimized through the use of site selection, design, construction, and maintenance practices appropriate to floodplain areas. In the United States, these practices are mainly based on the **National Flood Insurance Program (NFIP)** and its regulations governing land management and use (44 CFR Part 60).

The most effective flood mitigation solutions are relocation and elevation change, but when these methods are not cost effective or feasible, floodproofing may be an alternative solution.

The following are a few ways to achieve a flood-resistant building:

- Identify the nature, source, and severity of flood hazards affecting potential building sites.
- Select a building site to minimize or eliminate flood hazards.
- Determine any flood-related regulatory requirements. Check community (and state/regional) floodplain management regulations, applicable building code requirements, and for federal structures and facilities, Executive Order 11988 applies.

    Consensus standards are incorporated into building codes by reference; ASCE 7, *Minimum Design Loads for Buildings and Other Structures* (2010), and ASCE 24 are two consensus standards that are consistent with NFIP regulations. See section 2.1.7 of "Design Considerations in Floodproofing" published by FEMA for more information on consensus standards. Figure 3.11 shows how building design is regulated by the NFIP regulations and building codes.

- Determine present and future design flood characteristics anticipated at the selected site.
- Plan, design, and construct the building to minimize any potential flood damages by:

- elevating the building above the design flood level,
- designing the building foundation and any portions subject to flooding to withstand design flood conditions and loads,
- using flood-damage-resistant materials for any portions of the building below the design flood level, and
- where floodproofing is permitted, employing appropriate materials and methods to either dry-floodproof or wet-floodproof those portions of the building below the design flood level.

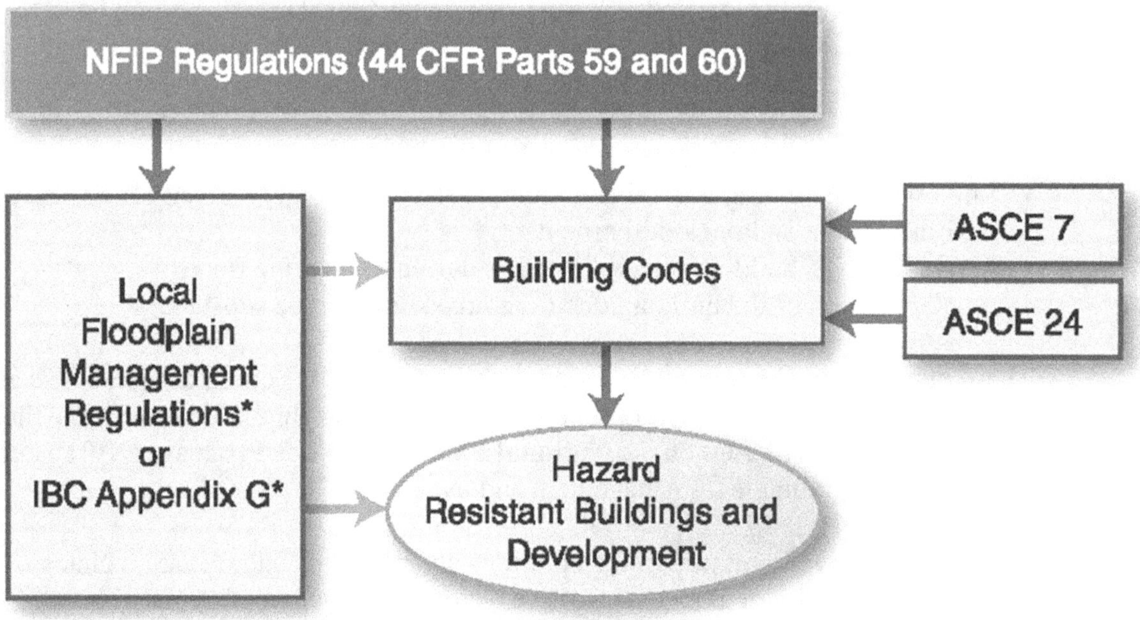

*NFIP-consistent administrative provisions, community-specific adoption of Flood Insurance Studies and maps, and technical requirements for development outside the scope of the building code (and higher standards, in some communities)

Figure 3.11 Satisfying NFIP requirements through building codes (Source: FEMA)

The agency that is most directly involved in identifying flood hazards (also, in the regulation of development within flood hazard areas, and in responding to flood disasters) is the Federal Emergency Management Agency (FEMA). Through its Mitigation Division, FEMA manages the National Flood Insurance Program (NFIP). Definitions and background for common flood hazard and mapping terms used below are available from FEMA.

The NFIP has mapped flood hazards in approximately 20,000 communities within the United States. Specifically, the NFIP has produced a series of maps— Flood Hazard Boundary Maps (FHBM), Flood Insurance Rate Maps (FIRM), and Flood Boundary and Floodway Maps (FBFM)—which depict special flood hazard areas (SFHA) and base flood elevations (BFEs). Flood Insurance Rate Maps (FIRMs—see sample below) are the most common of the flood hazard maps. Flood insurance study (FIS) reports are produced as part of each flood hazard study and provide data not shown on the FIRMs.

- **"A" zones** are those areas within the 100–year floodplain where high velocity wave action is not expected during a base flood. They can be mapped for inland or coastal floodplains.
- **"AE" zones** are the base floodplain where base flood elevations are provided. AE zones are now used on new format FIRMs instead of A1–A30 zones.
- **"V" zones** are those areas within a coastal floodplain where high velocity wave action can occur during a base flood. Base flood elevations (BFEs) are *not* determined. They are mapped exclusively in coastal floodplains.
- **"VE" zones** are the coastal areas subject to a velocity hazard (wave action) where BFEs are determined and provided on the FIRM.
- **"X" zones (unshaded)** are the areas determined to be outside the 500–year flood and protected by a levee from a 100–year flood.

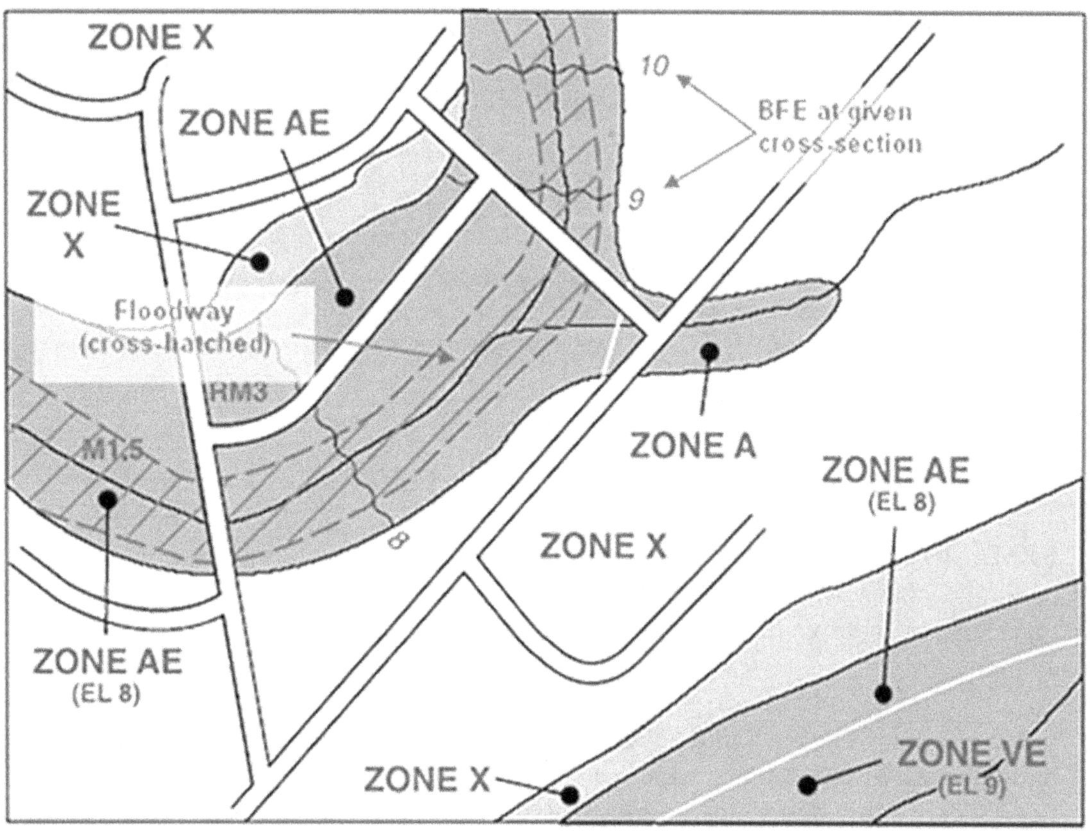

Figure 3.12 Sample riverine FIRM (Source: FEMA)

The base flood elevation (**BFE**) is the water surface elevation associated with the 100-year flood and has a 1 percent chance of being equaled or exceeded in any year.

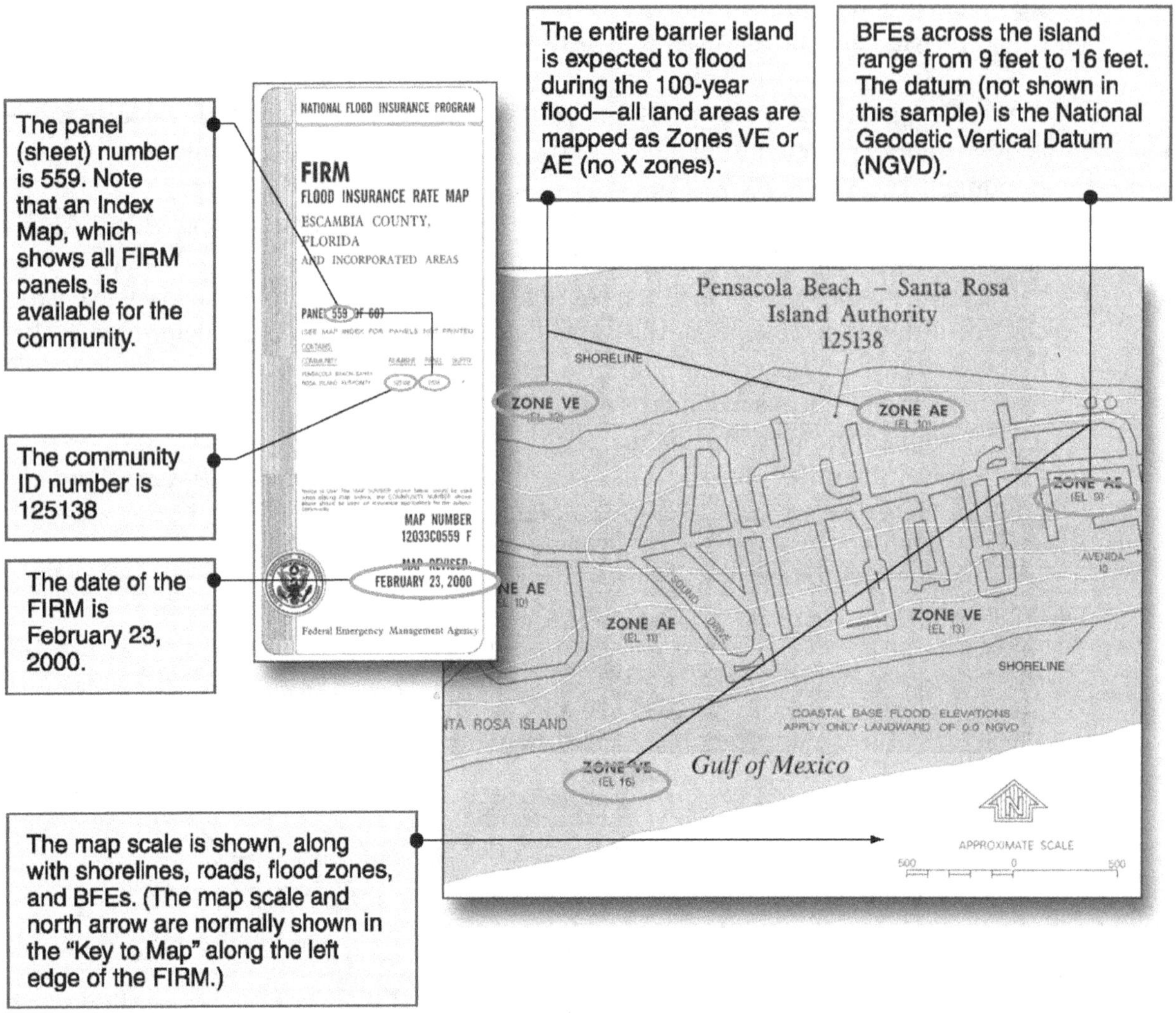

Figure 3.13 Sample coastal FIRM (Source: FEMA)

FEMA defines floodproofing as: "Any combination of structural and non-structural additions, changes, or adjustments to structures which reduce or eliminate flood damage to real estate or improve real property, water and sanitary facilities, structures and their contents."

Whole Building Design Guide (**WBDG**) defines **floodproofing** as the process of making a building resistant to flood damage, either by taking the building out of contact with floodwaters or by making the building resistant to any potential damage resulting from floodwaters. (Note that this discussion states, where permitted and used, that floodproofing must extend to the design flood elevation or **DFE**. In fact, **model building codes** and some floodplain management regulations require floodproofing to extend to levels above the DFE in certain instances. Designers should determine any specific floodproofing elevations for their projects, based on applicable regulations and code requirements.)

Note:
*A **model building code** is a building code developed and maintained by a standards organization independent of the jurisdiction responsible for enacting the building code.*

Some model building codes commonly used in the United States include:

- ***International Building Code (IBC)** by the International Code Council (**ICC**)*
- ***NFPA 5000** by the National Fire Protection Association (**NFPA**)*
- ***National Electrical Code** (**NEC**) by the National Fire Protection Association (**NFPA**)*

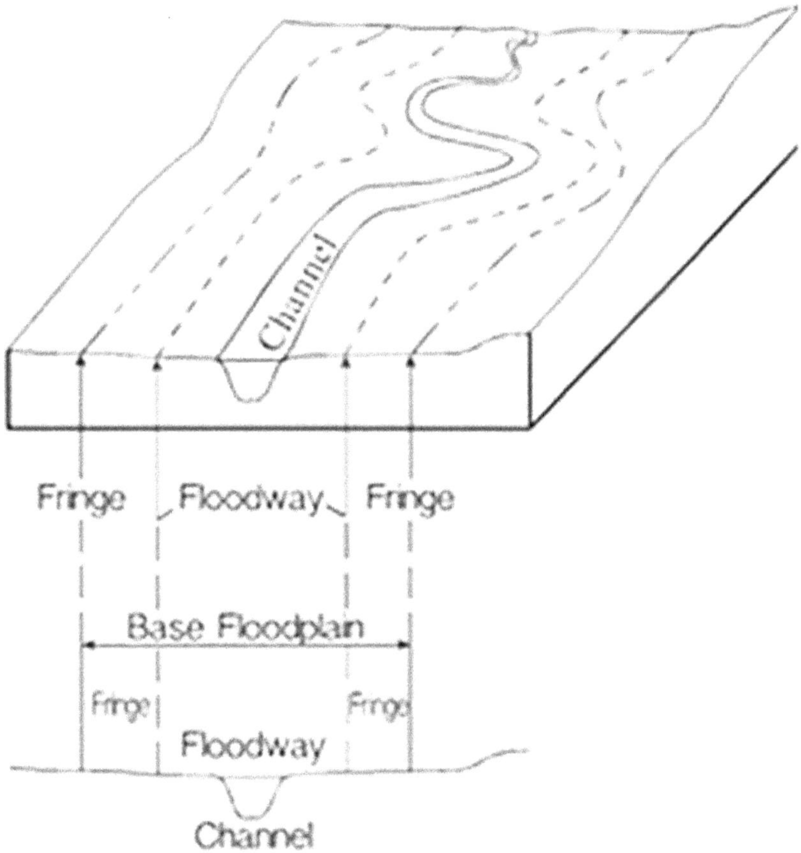

Figure 3.14 Relationship between a floodway and 100–year base floodplain (Source: FEMA)

Per WBDG, floodproofing can be subdivided into the following categories:

Dry-floodproofing requires use of special coatings, sealants, components and/or equipment to render the lower portion of a building substantially impermeable to the passage of water or watertight.

Wet-floodproofing allows the uninhabited lower portion of a building to flood, but uses materials that will not be damaged by flooding.

Active floodproofing, sometimes known as contingent (partial) or emergency (temporary) floodproofing, requires human intervention to implement actions that will protect a building and its contents from flooding. Successful use of this technique requires ample warning time to mobilize people and equipment and floodproofing materials.

Passive floodproofing, sometimes referred to as permanent floodproofing, requires no human intervention—the building (and/or its immediate surroundings) is designed and constructed to be floodproof without human intervention.

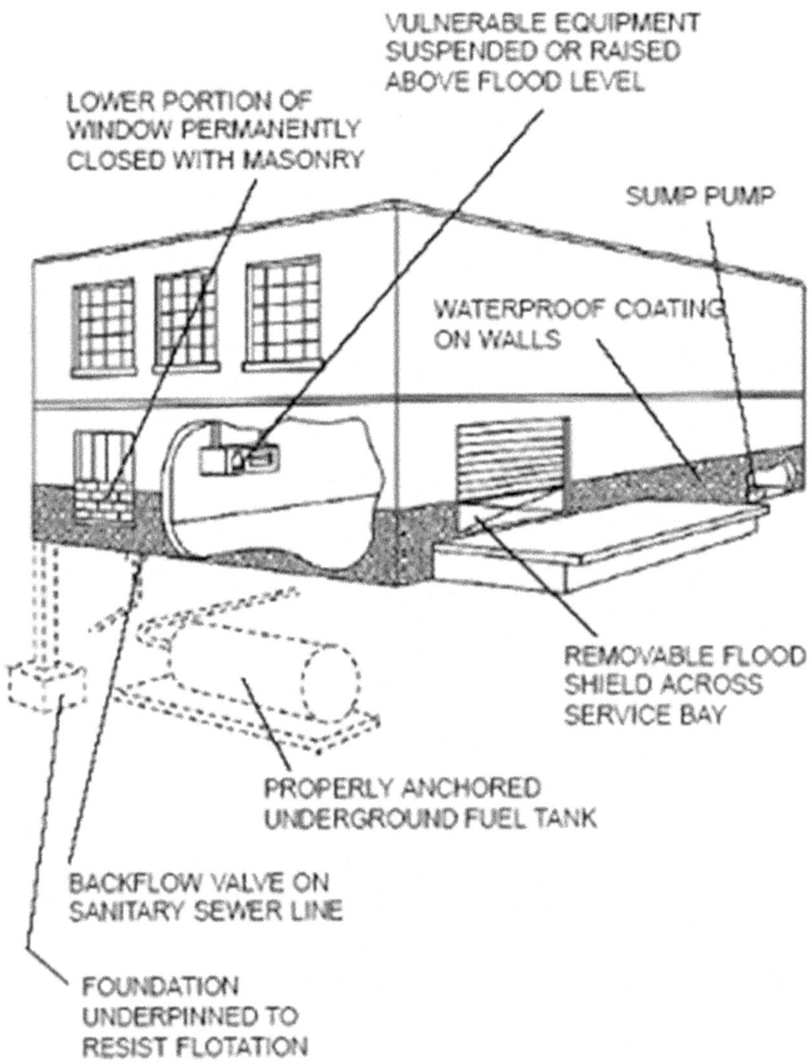

Figure 3.15 Typical dry-floodproofing techniques (Source: FEMA)

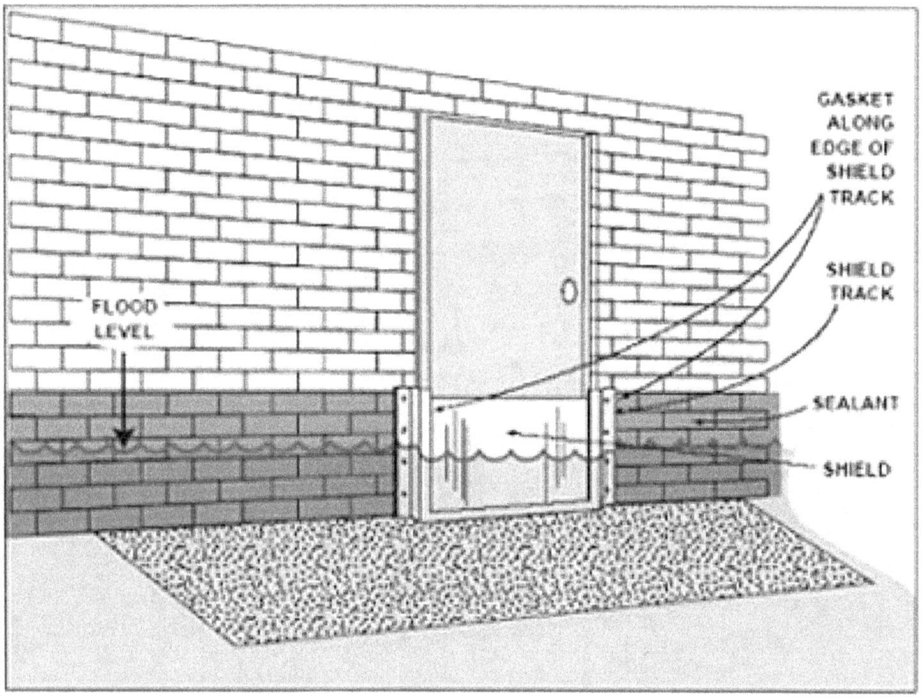

Figure 3.16 Temporary flood shield held in place by a permanent metal track (Source: FEMA)

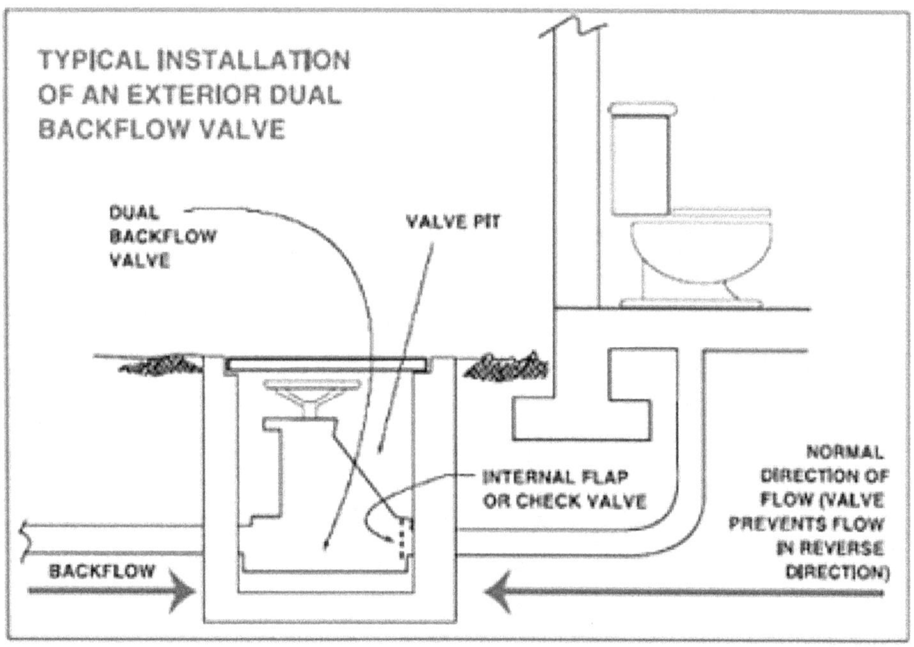

Figure 3.17 Backflow prevention valve (Source: FEMA)

Unless dry-floodproofed, enclosed areas below the DFE must be equipped with flood openings capable of equalizing water levels and hydrostatic loads (figure 3.18).

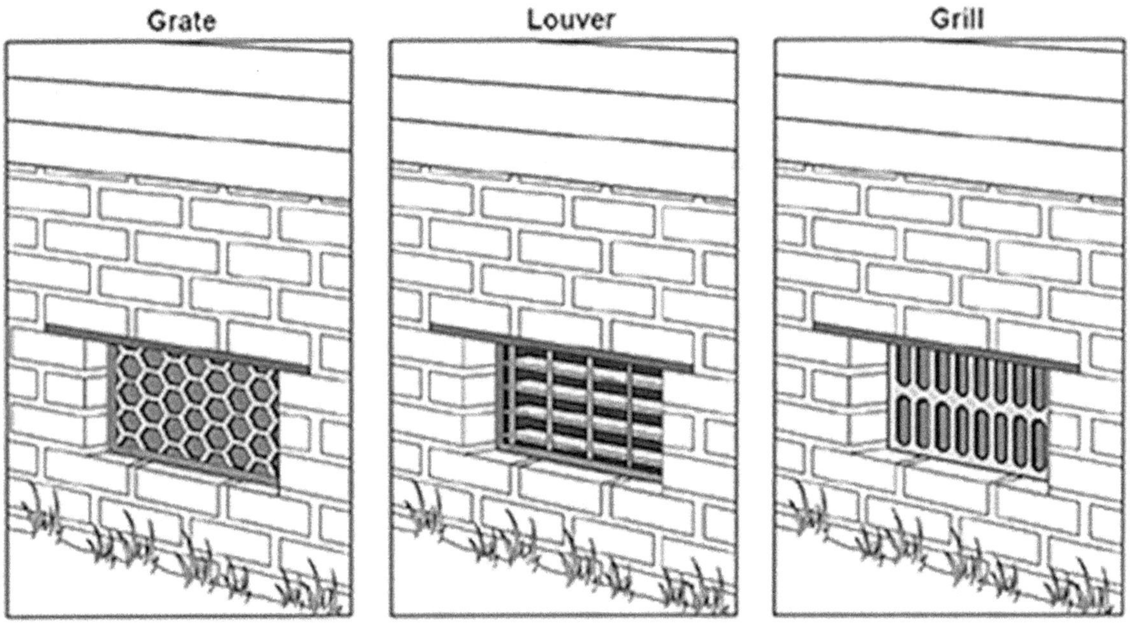

Figure 3.18 Typical flood opening covers (Source: FEMA)

Levees and floodwalls are barriers constructed away from a building as a method of preventing floodwaters.

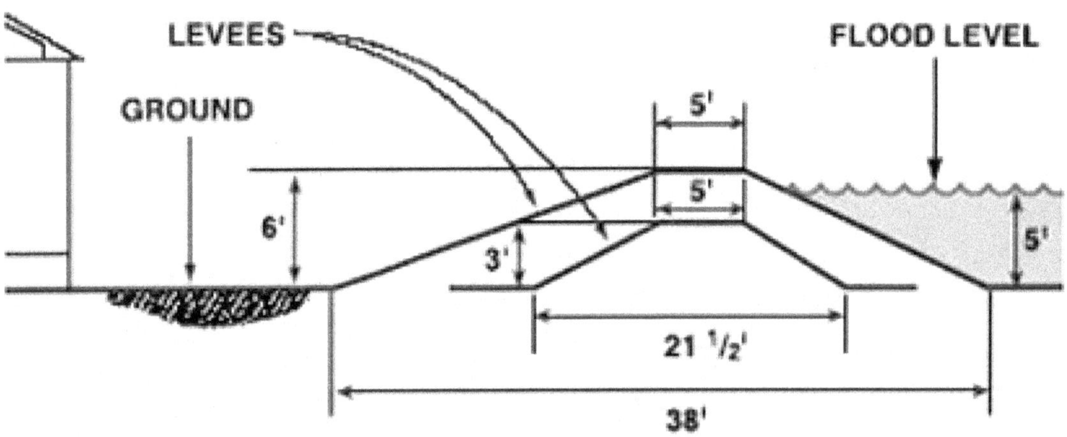

Figure 3.19 Cross-section of a typical three-foot-high and six-foot-high levee, showing the difference in the levee width (Source: FEMA)

Figure 3.20 Floodgate being installed to close a floodwall (Source: National Weather Service)

We have selected some very useful information from the PDF of "Design Considerations in Floodproofing" published by FEMA. This information is shown in figures 3.21-3.29.

Note:
You just need to look at the figures a few times to understand them. Do not memorize the information. This is typical for all the figures and diagrams in this book.

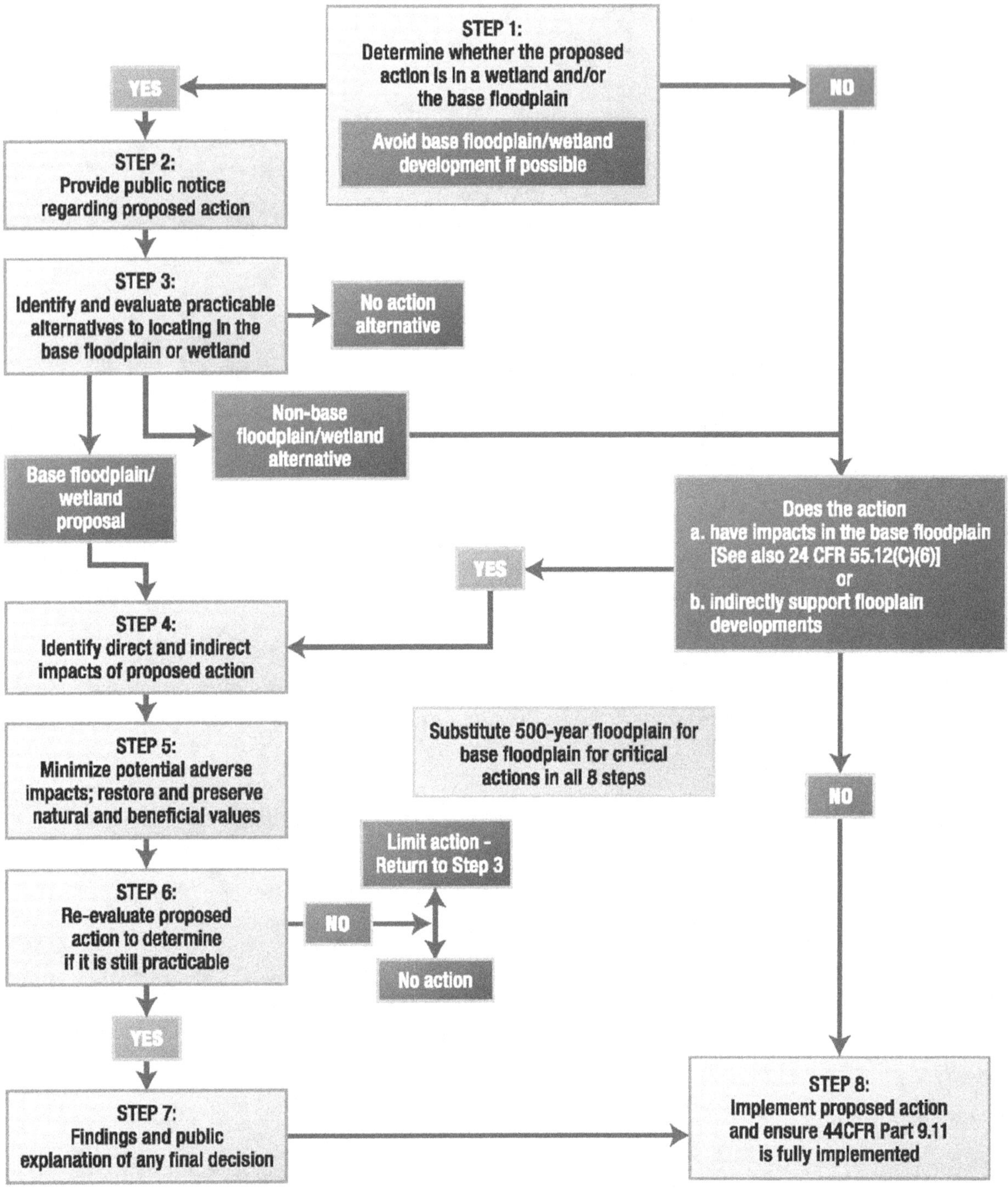

Figure 3.21 FEMA's eight-step decision-making process for compliance with Executive Order (EO) 11998 and 44 CFR Part 9 (Source: FEMA)

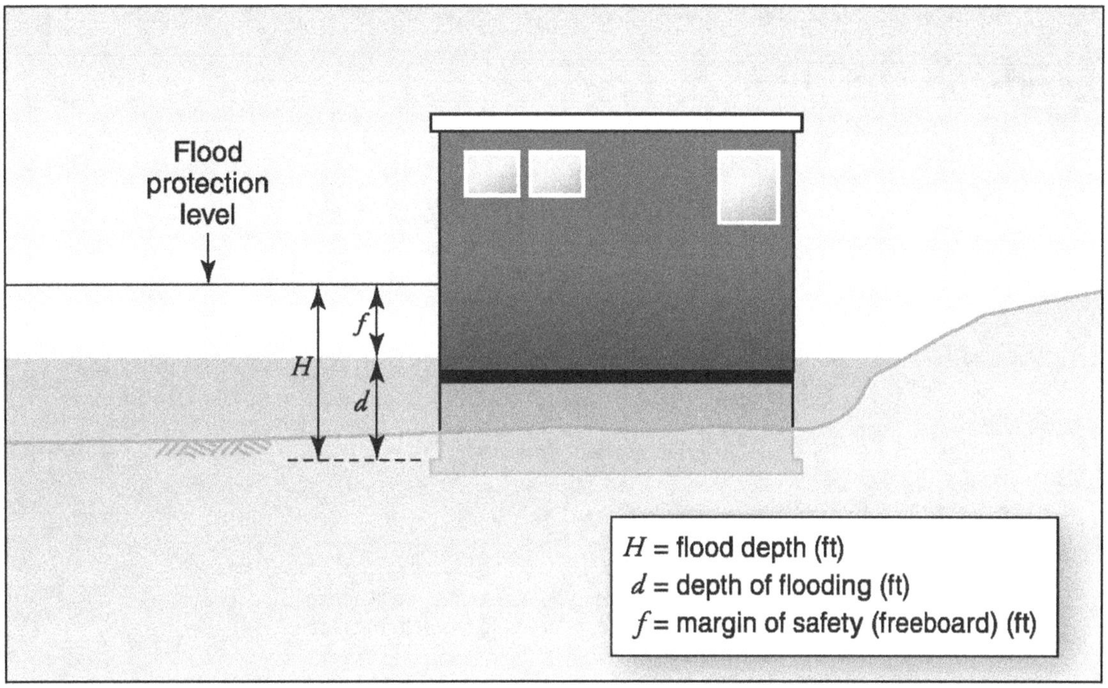

Figure 3.22 Flood depth (Source: FEMA)

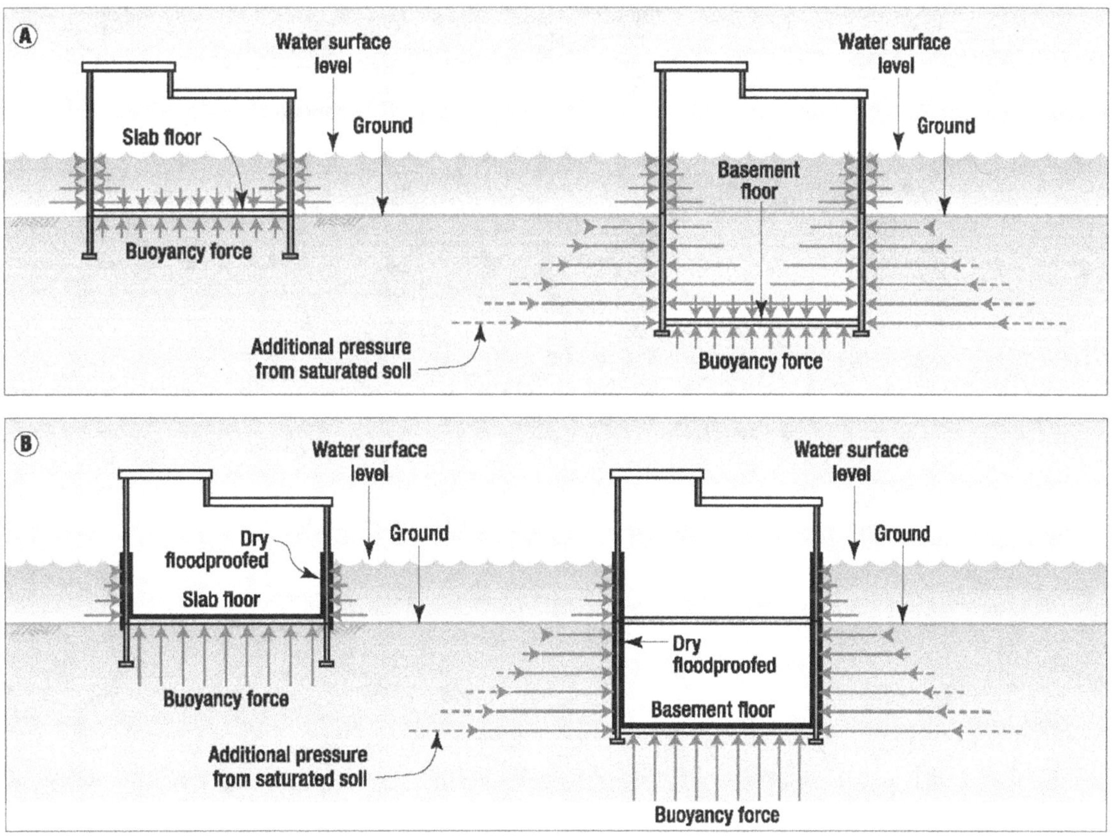

Figure 3.23 Equal (A) and unequal (B) hydrostatic pressures applied to the exterior elements of a building; (B) shows the building protected by dry floodproofing methods (Source: FEMA)

$$f_{dif} = \frac{1}{2}(S - \gamma_w) D^2$$

where:

f_{dif} = differential soil/water force acting at a distance $D/3$ from the point under consideration (lb/lf)

S = equivalent fluid weight of submerged soil and water (lb/ft^3) (shown in Table 4-3 of FEMA 259)

D = depth of saturated soil from adjacent grade to the (bottom of the flooded or submerged surface ft)

γ_w = specific weight of water (62.4 lb/ft^3 for fresh water and 64.0 lb/ft^3 for saltwater)

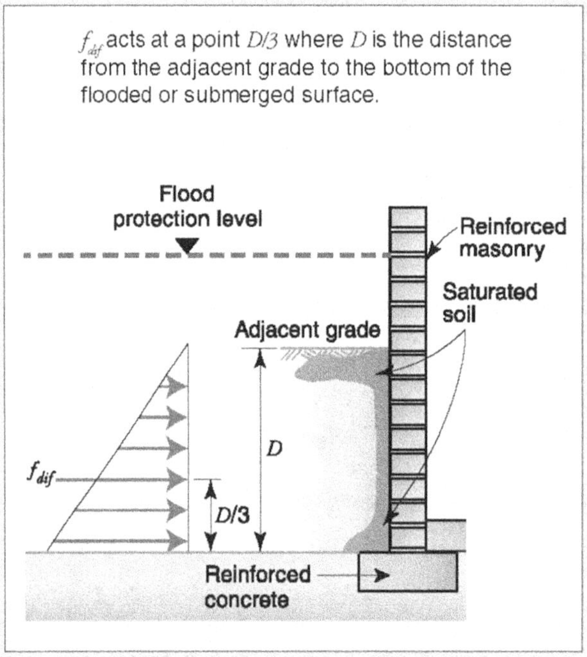

Figure 3.24 Submerged soil and water forces (Source: FEMA)

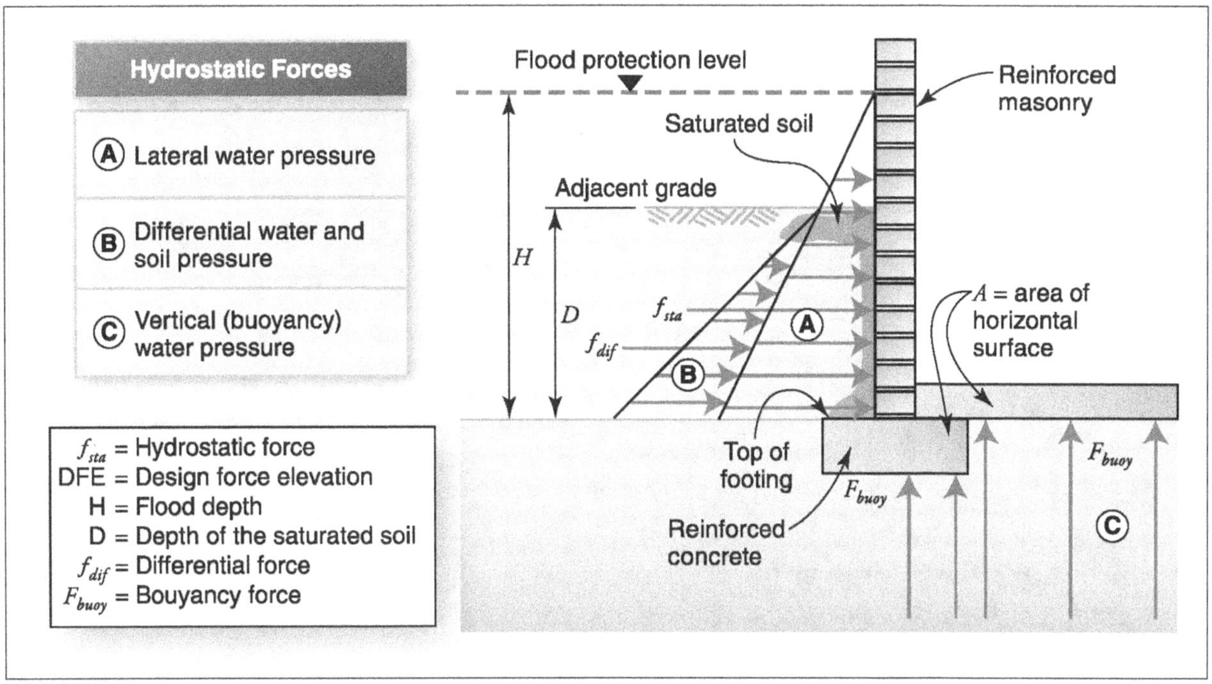

Figure 3.25 Hydrostatic forces acting on a foundation wall (Source: FEMA)

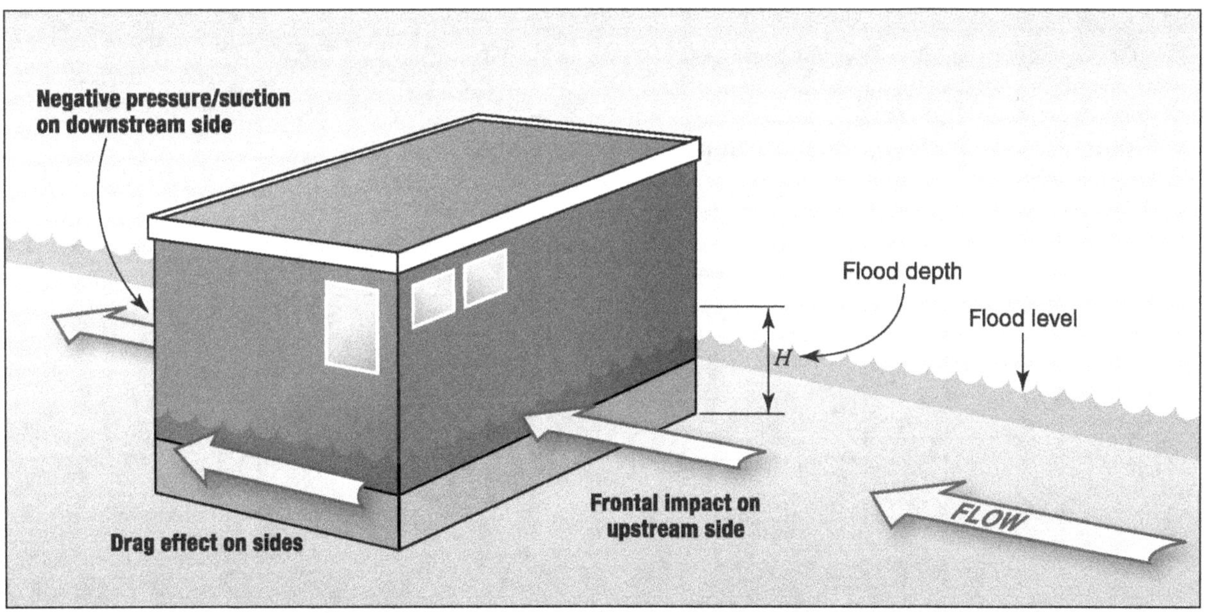

Figure 3.26 Hydrodynamic and impact loads (Source: FEMA)

$$F_{dyn} = C_d \rho \frac{V^2}{2} A$$

where:

F_{dyn} = horizontal drag force (lb)

C_d = drag coefficient (taken from Table 2-6)

ρ = mass density of fluid (1.94 slugs/ft^3 for fresh water and 1.99 slugs/ft^3 for saltwater)

V = velocity of floodwater (ft/sec)

A = surface area of obstruction normal to flow (ft^2) = $(w)(H)$ if the object is completely immersed

F_{dyn} acts at a point H/2, half way between the water elevation and the eroded ground surface

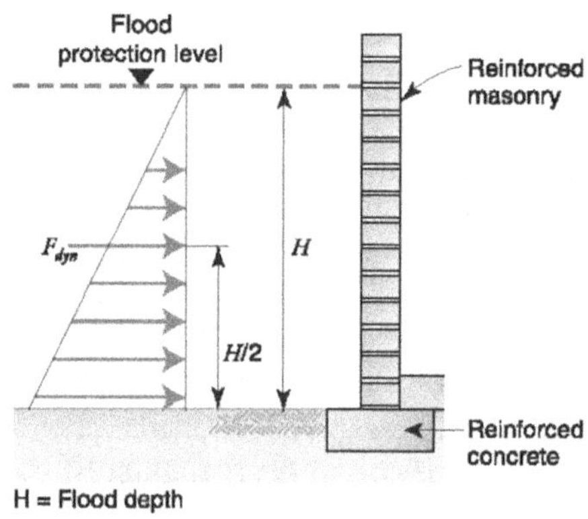

Figure 3.27 Hydrodynamic load (Source: FEMA)

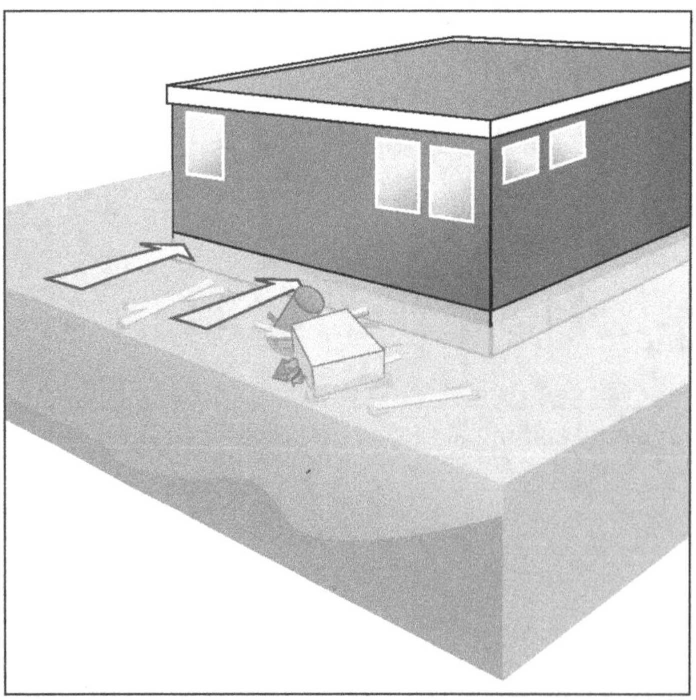

Figure 3.28 Frontal impact from debris (Source: FEMA)

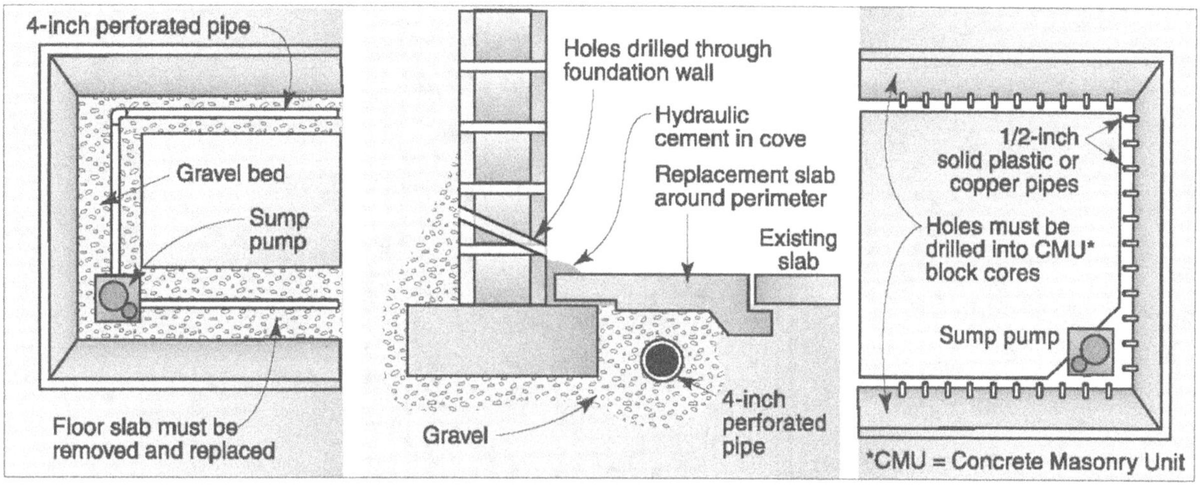

Figure 3.29 Typical interior drain systems (Source: FEMA)

Sump pumps are perhaps the most familiar method of dewatering below-grade areas. The sump is generally constructed so that its bottom is well below the base of the floor slab. Water in the areas adjacent to below-grade walls and floor migrate along the lines of least resistance, which should be toward and into the sump.

It may be necessary to provide a more readily accessible path of least resistance for water that has collected in the fill material and around the structure to follow. To achieve this, pipe segments are inserted and sometimes drilled through the below-grade wall and into the fill behind, *purposefully* allowing the interior to flood. Gravel is placed around the pipe segments on the exterior of the foundation wall to filter the surrounding soil, not allowing it to enter the interior of the building. The pipe segments are then connected to larger diameter pipes running along a gravel-filled trench or cove area into the floor slab and into one or more sumps (see Figure 3.29).

Interior drain systems can be overwhelmed by a quickly rising water table and are subject to potential power outages. (Source: FEMA)

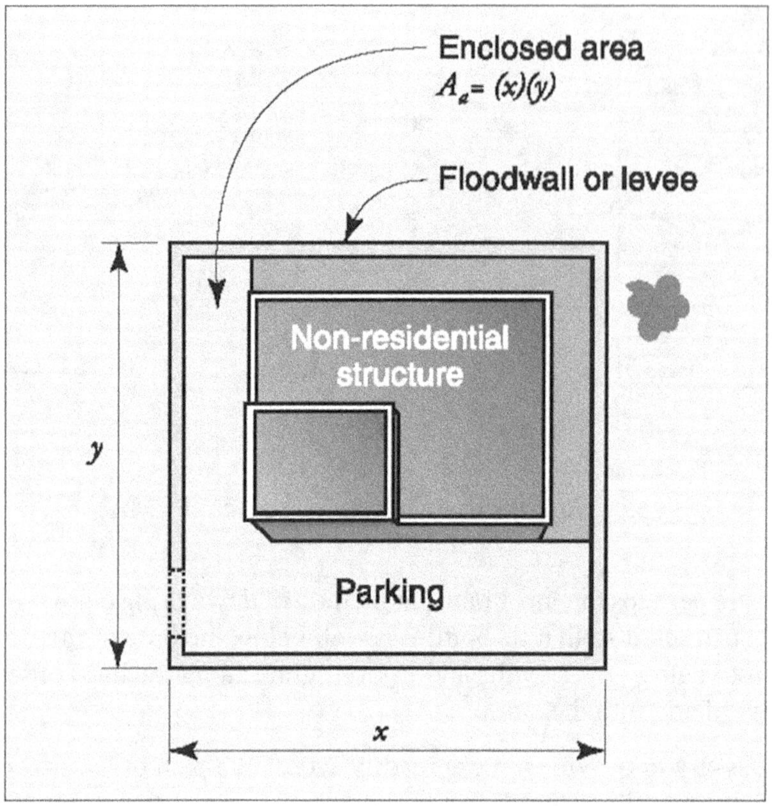

Figure 3.30 Rectangular area enclosed by a floodwall or levee (Source: FEMA)

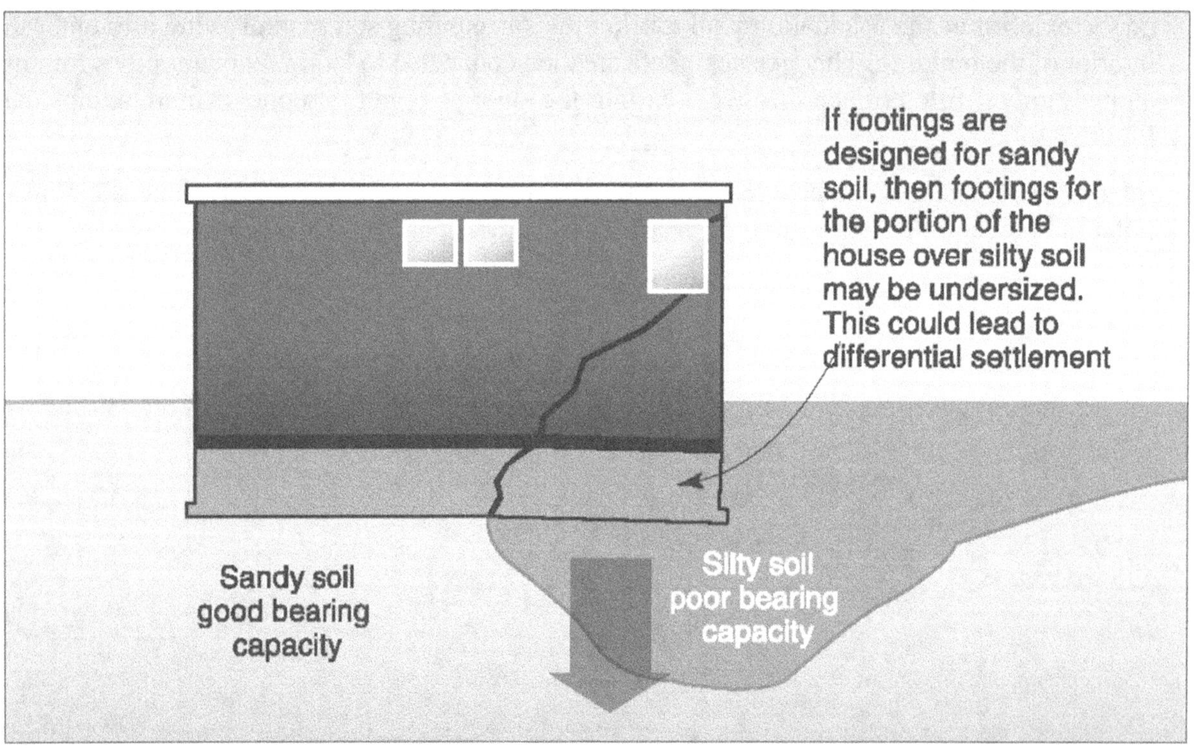

Figure 3.31 Significant differential settlement in the foundation causing slab failure

Flood frequency analyses define the probability that a flood of a specific size will be equaled or exceeded in any given year. A flood elevation with a 1 percent annual chance flood of being equaled or exceeded in a given year is referred to as the "base flood," commonly called the 100-year flood. Figure 3.32 illustrates the relationship between flood recurrence intervals and the probability of the event occurring within a given period. The probability that a base flood will be equaled or exceeded is 26 percent during a 30-year period. During a 70-year period (the potential useful life of many buildings), the probability increases to 50 percent. Although the base flood serves as the basis for NFIP insurance rates and regulatory floodplain management requirements, the relative frequency of any given flood (e.g., 2-year or 10-year) serves as a useful reference point when selecting a retrofitting option and evaluating cost effectiveness. (Source: FEMA)

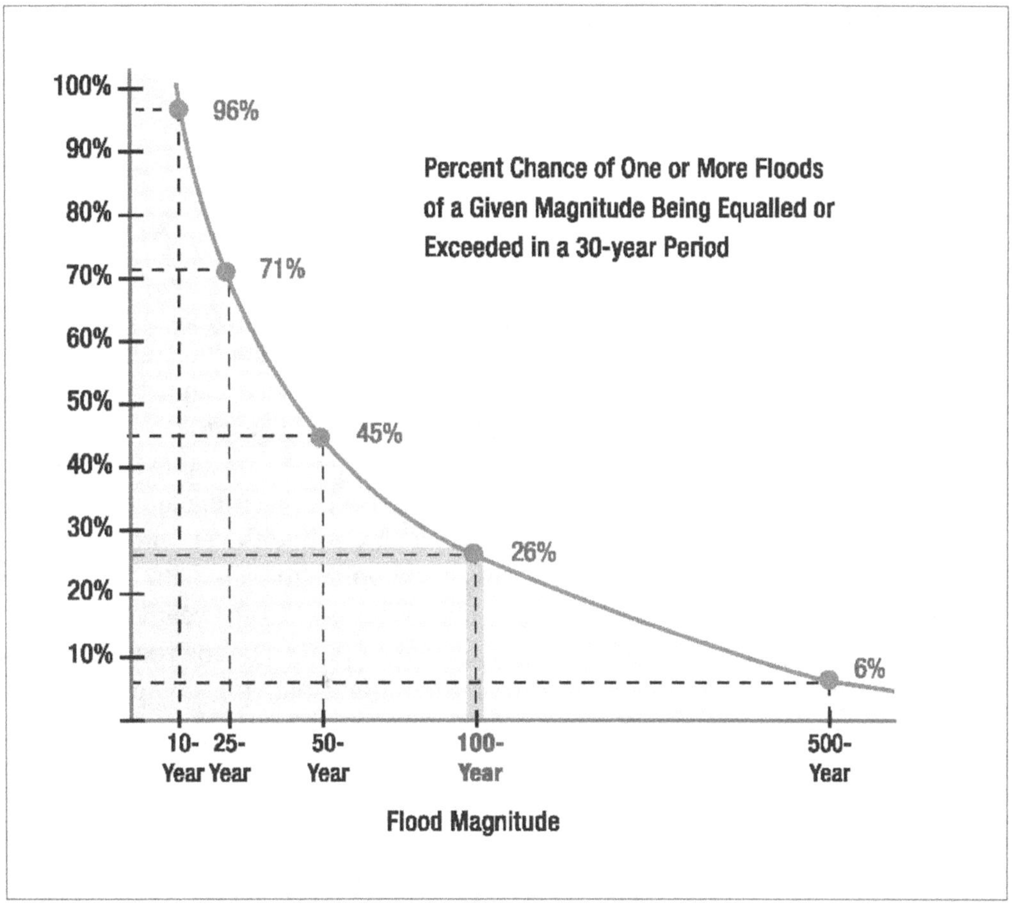

Figure 3.32 Relationship between flood recurrence intervals and the probability of an event occurring within a given period (Source: FEMA)

The information we provide in this book should be adequate for you to prepare for the ARE on this subject. However, if you want to know more about this subject and read detailed discussions, you can download the free PDF of "Design Considerations in Floodproofing" published by FEMA at link below:
http://www.fema.gov › media-library-data › P-936_sec2_508

Notes:
You can download the aforementioned free PDF, and at least look through it two or three times. Do not read the words, just look at the diagrams and figures. If you already understand them, then move on to the next page. If you do not understand, then stop and read the associated text. This approach will help you cut out many pages of reading.

"Flood Resistance of the Building Envelope" written by Christopher P. Jones, PE, provides a lot of useful information. You can find this article at the following link:
https://www.wbdg.org/resources/flood-resistance-building-envelope

Notes:
Look through this article two or three times. Do not read the words, just look at the diagrams and figures. If you already understand them, then move on to the next page. If you do not understand, then stop and read the associated text. This will help you cut out many pages of reading.

2. **Unstable Soil**
Unstable soil is soil that will not stay in place on its own, and requires extra support.

The following are the main causes for unstable soil:
- **Erosion** is processes in which external elements (wind, water, etc.) remove rock or soil from one place and move it to another. There are many different erosion types, including gull and river erosion, wind erosion, and erosion caused by human activity. Erosion destabilizes soil and can lead to sinkholes and landslides.
- **Poor compaction** is one of the most common causes of unstable soil. Some types of soil are very loose and hard to compact, often caused by an imbalance of organic matter, mineral pieces, air, and water. For instance, a clay soil with very high moisture content is unstable and difficult to compact. This also applies to soils with high sand content.
- **Freezing and thawing** accelerate erosion processes. Cold weather freezes moisture trapped in tiny cracks. When this water freezes, it pushes on the rocks and breaks them into smaller pieces. The processes of freezing and thawing continually break down rocks and sediments.
- **Decomposition** of organic materials in soil causes rapid changes in mass and form. When topsoil contains a high concentration of organic materials, this material will decompose and become unstable. Up to 90 percent of organic material will disappear in the decomposition process.

Solutions for unstable soils include the following:
- Use drainage tile to drain excess moisture out of the soils.
- Tilt and turn over the soil to help dry it.
- Excavate into the dirt subgrade and replace the soil with a proper base material to help bridge the unstable soils.
- Compact the soils.
- Use ultra-low viscosity polyurethane resins and foaming and permeation polyurethanes to restore soil stability.
- Use chemical treatments.

- Use foamed asphalt injection to strengthen the base.
- Utilize **piles** for support without removing soils. Piles include end bearing piles or friction piles.

3. **Radon**

Per EPA, **Radon** is a naturally occurring radioactive gas that can cause lung cancer. You can't see or smell radon. Exposure to radon causes lung cancer in non-smokers and smokers alike. Testing is the only way to know your level of exposure.

Where can you get a radon test kit? Per EPA, there are a few ways:

- The National Radon Program Services at Kansas State University offers discounted test kits available to purchase online. You may complete the test kit order form electronically and print it out to mail or fax in. You may download a radon test kit coupon and mail it in. See **National Radon Program Services** at the following link: https://sosradon.org/test-kits

- Many radon test kits can be found online or in home improvement stores. Follow the directions on the packaging for the proper placement of the device and where to send the device after the test to find out your radon level.

- Contact your **state radon program** for information on how to obtain a test kit from a radon measurement professional. Some states also offer free or discounted test kits to the public. See **state radon program** at the following link: https://www.epa.gov/radon/find-information-about-local-radon-zones-and-state-contact-information

You can also find "EPA Map of Radon Zones including State Radon Information and Contacts" at afore-mentioned link. The radon zones are color-coded:
Zone 1: Counties with predicted average indoor radon screening levels greater than 4 pCi/L
Zone 2: Counties with predicted average indoor radon screening levels from 2 to 4 pCi/L
Zone 3: Counties with predicted average indoor radon screening levels less than 2 pCi/L

Note:
*Picocuries per liter (**pCi/L**) is a unit for measuring radioactive concentrations. The curie (Ci) unit is the activity of 1 gram of pure radium 226. Pico is a scientific notation term which means 10^{-12}. A **picocurie** is one trillionth of a curie. Thus, a **picocurie** (abbreviated as pCi) represents 2.2 disintegrations per minute.*

Radon is found in soil and rock. It seeps into homes through cracks in the foundation, walls and joints.

New buildings can be built to resist radon entry. The additional cost at the time of construction is minimal. Consider these facts:

- For a builder, it is much less expensive to install a radon-resistant system during construction than to go back and fix a radon problem identified later.
- If a new homeowner tests for radon and has to mitigate high levels, it could cost the builder or the owner more than an initial installation.

EPA recommends homes be fixed if the radon level is 4 pCi/L or more. Because there is no known safe level of exposure to radon, EPA also recommends that Americans consider fixing their home for radon levels between 2 pCi/L and 4 pCi/L.

The average indoor radon concentration for America's homes is about 1.3 pCi/L. It is upon this national average indoor level that EPA based its estimate of 21,000 radon-related lung cancers a year. The average concentration of radon in outdoor air is .4 pCi/L or 1/10th of EPA's 4 pCi/L action level.

Mitigation of radon in the air is achieved through ventilation, either collected below a concrete floor slab or in a membrane on the ground, or by increasing the air changes per hour in the building. Treatment systems using activated charcoal or aeration are available to remove radon from domestic water supplies.

EPA generally recommends methods that prevent the entry of radon. Soil suction, for example, prevents radon from entering your home by drawing the radon from below the home and venting it through a pipe, or pipes, to the air above the home where it is quickly diluted.

For projects that require radon mitigation, the management or construction team shall have at least one professional with special radon mitigation training.

Figure 3.33 shows a passive radon control systems for new construction (Source: FEMA)

See EPA website for more information:
https://www.epa.gov

4. **Lead**

Per EPA, lead is a naturally occurring element found in small amounts in the earth's crust. While it has some beneficial uses, it can be toxic to humans and animals, causing health effects.

Lead can be found in all parts of our environment – the air, the soil, the water, and even inside our homes. Much of our exposure comes from human activities including the use of fossil fuels including past use of leaded gasoline, some types of industrial facilities, and past use of lead-based paint in homes. Lead and lead compounds have been used in a wide variety of products found in and around our homes, including paint, ceramics, pipes and plumbing materials, solders, gasoline, batteries, ammunition, and cosmetics.

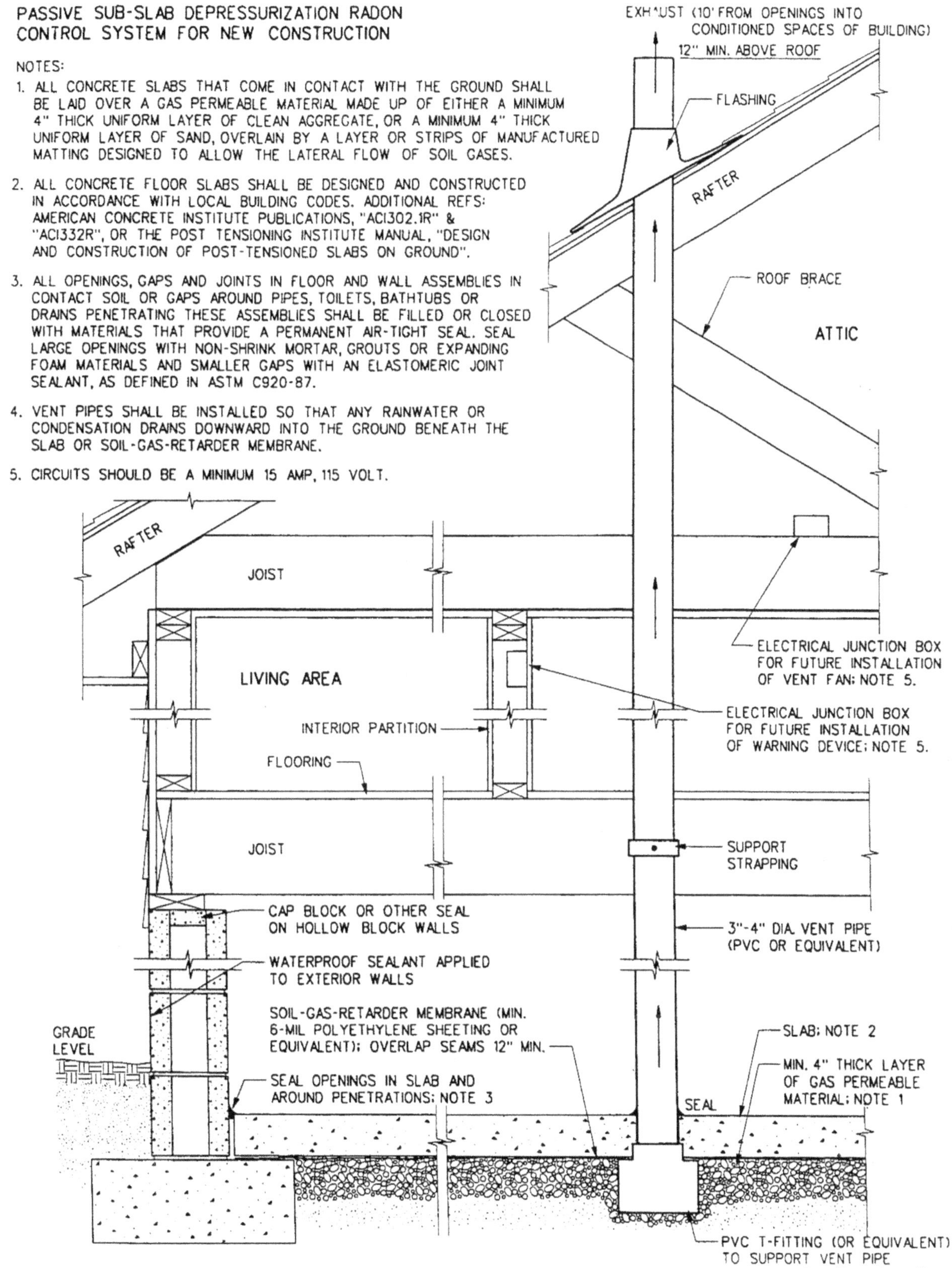

Figure 3.33 Passive radon control systems for new construction (Source: FEMA)

Lead may enter the environment from these past and current uses. Lead can also be emitted into the environment from industrial sources and contaminated sites, such as former lead smelters. While natural levels of lead in soil range between 50 and 400 parts per million, mining, smelting, and refining activities have resulted in substantial increases in lead levels in the environment, especially near mining and smelting sites.

When lead is released to the air from industrial sources or spark-ignition engine aircraft, it may travel long distances before settling to the ground, where it usually sticks to soil particles. Lead may move from soil into ground water depending on the type of lead compound and the characteristics of the soil.

Federal and state regulatory standards have helped to reduce the amount of lead in air, drinking water, soil, consumer products, food, and occupational settings.

Learn more about sources of lead exposure:

- **At home:** If your home was built before 1978, there is a good chance it has lead-based paint. In 1978, the federal government banned consumer uses of lead-containing paint, but some states banned it even earlier. Lead from paint, including lead-contaminated dust, is one of the most common causes of lead poisoning.
- **At schools and childcare facilities:** Although your home may be free of lead-based paint hazards, your child could still be exposed elsewhere, particularly if they spend time in a building built before 1978. Find out if your child has elevated levels of lead in his or her blood. Because lead poisoning often occurs with no obvious symptoms, it frequently goes unrecognized. You can test your child for lead poisoning by asking your pediatrician to do a simple blood test. Children with elevated blood lead levels can have serious health effects.
- **In products:** Lead can also be found in painted toys, furniture and toy jewelry, cosmetics, food or liquid containers, plumbing products.
- **In drinking water:** Lead can enter drinking water through corrosion of plumbing materials, especially where the water has high acidity or low mineral content that corrodes pipes and fixtures. Homes built before 1986 are more likely to have lead pipes, fixtures and solder. However, new homes are also at risk: even legally "lead-free" plumbing may contain up to 8 percent lead.
- **In outdoor air:** Sources of lead emissions vary from one area to another. EPA has a national database that shows which types of sources contribute to the total lead emitted into the air each year on a national, state, and local level. At the national level, major sources of lead in the air are ore and metals processing and piston-engine aircraft operating on leaded aviation fuel. Other sources are waste incinerators, utilities, and lead-acid battery manufacturers. The highest air concentrations of lead are usually found near lead smelters.
- **In soil:** Lead is naturally-occurring, and it can be found in high concentrations in some areas. In addition, soil, yards and playgrounds can become contaminated when exterior lead-based paint from houses or buildings flakes or peels and gets into the soil. Soil may also be contaminated from past use of leaded gasoline in cars, from industrial sources, or even from contaminated sites, including former lead smelters.

- **In dust:** lead in household dust results from indoor sources such as old lead paint on surfaces that are frequently in motion or bump or rub together (such as window frames), deteriorating old lead paint on any surface, home repair activities, tracking lead contaminated soil from the outdoors into the indoor environment, or even from lead dust on clothing worn at a job site.

Simple steps like keeping your home clean and well-maintained will go a long way in preventing lead exposure. You can lower the chances of exposure to lead in your home, both now and in the future, by taking these steps:

- Inspect and maintain all painted surfaces to prevent paint deterioration
- Address water damage quickly and completely
- Keep your home clean and dust-free
- Clean around painted areas where friction can generate dust, such as doors, windows, and drawers. Wipe these areas with a wet sponge or rag to remove paint chips or dust
- Use only cold water to prepare food and drinks
- Flush water outlets used for drinking or food preparation
- Clean debris out of outlet screens or faucet aerators on a regular basis
- Wash children's hands, bottles, pacifiers and toys often
- Teach children to wipe and remove their shoes and wash hands after playing outdoors
- Ensure that your family members eat well-balanced meals. Children with healthy diets absorb less lead.
- If you are having home renovation, repairs, or painting done, make sure your contractor is Lead-Safe Certified, and make sure they follow "lead safe work practices" (PDF). See link:
  https://www.epa.gov/lead/lead-safe-certified-guide-renovate-right

5. **Brownfields**

Per EPA, a **brownfield** is a property, the expansion, redevelopment, or reuse of which may be complicated by the presence or potential presence of a hazardous substance, pollutant, or contaminant. It is estimated that there are more than 450,000 brownfields in the U.S. Cleaning up and reinvesting in these properties increases local tax bases, facilitates job growth, utilizes existing infrastructure, takes development pressures off of undeveloped, open land, and both improves and protects the environment.

EPA's "Anatomy of Brownfields Redevelopment" is a guide that explains the real estate process for cleaning up and redeveloping one or more brownfields sites, including:

- Opportunities and Risks Specific to Brownfields Redevelopment
- Typical Brownfields Redevelopment Scenarios: Private-Led Brownfields Redevelopment; Public-Led Brownfields Redevelopment; Public-Private Partnership Brownfields Redevelopment
- Three Stages of the Brownfields Redevelopment Process: pre-development, development, and management
- Paying for Brownfields Redevelopment
- Critical Players in Brownfields Real Estate Development
- Two Brownfields Redevelopment Case Studies

You should download "Anatomy of Brownfields Redevelopment" and read it two or three times. You can download it at the following link:
https://www.epa.gov/brownfields/anatomy-brownfields-redevelopment-0

EPA's Brownfields Program is designed to empower states, communities, and other stakeholders in economic redevelopment to work together in a timely manner to prevent, assess, safely clean up, and sustainably reuse brownfields. Beginning in the mid-1990s, EPA provided small amounts of seed money to local governments that launched hundreds of two-year Brownfields pilot projects and developed guidance and tools to help states, communities and other stakeholders in the cleanup and redevelopment of brownfields sites. The 2002 Small Business Liability Relief and Brownfields Revitalization Act codified many of EPA's practices, policies and guidance. This expanded EPA's assistance by providing new tools for the public and private sectors to promote sustainable brownfields cleanup and reuse. The 2018 Brownfields Utilization, Investment and Local Development (BUILD) Act reauthorized EPA's Brownfields Program, and authorized changes that affect brownfield grants, ownership and liability provisions, and State & Tribal Response Programs.

Brownfield grants continue to serve as the foundation of EPA's Brownfields Program. These grants support revitalization efforts by funding environmental assessment, cleanup, and job training activities.

See the following link for more information:
https://www.epa.gov/brownfields/overview-epas-brownfields-program

Brownfield remediation is the sealing off or removal of that contaminant so that a site can be reused without health concerns.

Protect teams can use the American Society for Testing and Materials (ASTM) **Environmental Site Assessment (ESA)** methodology for identifying and investigating a site's environmental contamination. This prerequisite encourages the use of an ESA (or local equivalent) and remediation of any confirmed site contamination to protect human health. Local assessment standards are acceptable if they are at least as stringent as ASTM Phase I and II ESA.

Green Building Alliance has provided some valuable information on brownfield remediation history, process and methods, and includes remediation techniques which are divided into those for groundwater and those for soil. This information is available at the following link:
https://www.go-gba.org/resources/green-building-methods/brownfield-remediation/

6. **Abandoned structures**
Abandoned structures are unoccupied structures in a state of grave disrepair, perhaps strewn with trash, boarded up, and scrawled with graffiti.

Reusing and revitalizing abandoned structures is an important aspect of sustainability which can improve our communities and environment, reduce crimes, avoid unnecessary demolition, and reduce waste going to the landfill.

Adaptive reuse is a process of retrofitting old buildings for new uses and retaining historic structural integrity while meeting current needs.

There are many societal benefits to adaptively reusing our buildings. Vacant lots and abandoned buildings drive down property values, create a sense of hopelessness and economic decline, and invite crime. Reusing abandoned structures maintains the character of our cities and strengthens our civic pride.

There can be many cost advantages to reusing an older or abandoned structure, such as lower establishment costs. Further, there is little to no demolition required, land acquisition often costs less, and many, if not all, of the required services and utilities are already connected and may only need modernization. Also, there are additional savings because the structure is already built.

Reusing abandoned structures may help you to achieve points in the LEED (Leadership in Energy and Environmental Design) building rating system.

7. **Other natural or built features**
Other natural or built features can have significant impacts on architectural designs. For example, at a site with steep slopes, stepped footings can reduce a lot of cut and fill resulting in lower construction costs and a unique design.

C. Determine optimal use of onsite resources by incorporating sustainability principles (U/A)

NCARB Requirements
After evaluating a site for both opportunities and constraints, you must be able to identify appropriate design responses for a building's location, orientation, massing, footprint, and passive energy use in order to minimize the project's negative environmental impacts and make use of existing natural features.

See our discussions on sun, wind, water, topography, and temperature at Section A of this chapter.

1. **LEED rating systems and references guides**
Sustainable building design, or green building, has become more and more important in architectural design. LEED (Leadership in Energy and Environmental Design) is a major green building rating system developed by the U.S. Green Building Council (USGBC). It turns the abstract concepts of green building and sustainability into something that can be quantified and fine-tuned. This quantification is a major development in the green building movement.

LEED serves as a bridge between professionals and ordinary people. The professionals can dive into the specific LEED rating system and know what can achieve points to make a business or building green, while ordinary people just need to know the results, i.e., a building achieves LEED Platinum, LEED Gold, LEED Silver, or LEED certified rating. This is similar to the Olympic medal system. The referee knows all the technical stuff and how to score each performance, while ordinary people just need to know the results or who gets a gold, silver, or bronze medal.

LEED is one of the most important trends in development and is revolutionizing the construction industry. It has gained tremendous momentum and has a profound impact on our environment. I have published a number of books on LEED, and you can find them at the following link:
http://www.GreenExamEducation.com/

The main purpose of the USGBC are attempting to improve the way a building is designed, built, and used in order to achieve a healthy, profitable, and environmentally responsible building and environment. In turn this is meant to improve quality of life.

The USGBC emphasizes not only the decisions themselves but also *how* the decisions are made.

The guiding principles of the USGBC are:

- advocate the **triple bottom lines** (to balance environmental, social, and economic needs or sometimes summarized as planet, people, and profit)
- build leadership
- strive to achieve the balance between humanity and nature
- uphold integrity and restore, preserve, and protect the environment, species, and ecosystem
- use democratic and interdisciplinary approaches to ensure inclusiveness to achieve a common goal; and openness, honesty, and transparency

The USGBC uses the following areas to measure the performance of a building's sustainability:

Integrative Process (IP), Location and Transportation (LT), Sustainable Sites (SS), Water Efficiency (WE), Energy and Atmosphere (EA), Materials and Resources (MR), Indoor Environmental Quality (EQ), Innovation (IN), and Regional Priority (RP).

Mnemonic: Ian and Larry, Shall We Eventually Make Everything Italic and Red?

The USGBC has the following reference guides and green rating system portfolios.

1) *The LEED Reference Guide for Building Design and Construction v4 (BD&C)* covers the following LEED rating systems:
   - LEED BD+C: New Construction
   - LEED BD+C: Core and Shell Development
   - LEED BD+C: Schools
   - LEED BD+C: Retails
   - LEED BD+C: Data Center

- LEED BD+C: Warehouse and Distribution Center
- LEED BD+C: Hospitality
- LEED BD+C: Health Care
- LEED BD+C: Homes
- LEED BD+C: Multi-family Midrise

2) *The LEED Reference Guide for Interior Design and Construction v4 (ID&C)* covers the following LEED rating systems:
   - LEED ID+C: Commercial Interiors
   - LEED ID+C: Retails
   - LEED ID+C: Hospitality

3) *The LEED Reference Guide for Green Building Operations and Maintenance v4 (O&M)* covers the following LEED rating systems:
   - LEED O+M: Existing Buildings
   - LEED O+M: Retails
   - LEED O+M: Schools
   - LEED O+M: Hospitality
   - LEED O+M: Data Center
   - LEED O+M: Warehouse and Distribution Center

   Note:
   These LEED-EB rating systems are the only systems that cover building operation. All other LEED systems cover building design and construction, but NOT operation.

4) *The LEED for Neighborhood Development Reference Guide v4 (LEED-ND)* covers the following:
   - LEED ND: Plan
   - LEED ND: Built Project

5) **Campus Guide**. See the link below:
   http://www.usgbc.org/campusguidance

2. **Association factors in LEED rating systems**
   Credits that can contribute to LEED's **"Impact Categories"** are given more points. These impact categories are weighted through a consensus driven process and are as follows:
   - **global climate change (35%)**
   - **social equity, environmental justice, and community quality of life (5%)**
   - **individual human health and well-being (20%)**
   - **greener economy (5%)**
   - **biodiversity and Ecosystem (10%)**
   - **water resources (15%)**
   - **sustainable and regenerative material resources cycles (10%)**

The USGBC uses three **association factors** to measure and scale credit outcome to a given impact category component.

1) **Relative efficacy** measures whether a credit outcome has a positive or negative association with a given Impact Category component, and how strong that association is. They are rated as follows:
   - no association
   - low association
   - medium association
   - high association
   - negative association

2) **Benefit duration** measures how long the benefits or consequences of the credit outcome will last:
   - 1-3 Years
   - 4-10 Years
   - 11-30 Years
   - 30+ Years (Building/Community Lifetime)

3) **Controllability of effect** indicates which individual is most directly responsible for achieving the expected credit outcome. The more a credit outcome depends on active human effort, the less likely it will be achieved with certainty, and the credit will have fewer points. Less human effort equals more points.

The USGBC simplifies the weighting process of points into a **scorecard computed as follows**:

- **100 base points** for the base LEED Rating System
- **1 point minimum** for each credit
- **whole points** and no fractions for LEED points

The USGBC has four levels of building certification. The four levels and the points required for each are as follows:

Certified	40–49	points
Silver	50–59	points
Gold	60–79	points
Platinum	80	points and above

The level achieved by a building is based on the number of points earned under the LEED green building rating system. The points are calculated using the scorecards or checklists provided by the USGBC.

No matter which **LEED v4.0** rating system you choose, each LEED v4.0 rating system has **100** base points; a maximum of **6** possible extra points for Innovation (IN), and a maximum of **4** possible extra points for Regional Priority (RP) for each project. There are 6 possible RP points, but you can only pick and choose a maximum of 4 points for each project.

Refer to the following link for the **Regional Priority Credits (RPC)** for all 50 States. With this spreadsheet, you can locate the RPC for your area by zip code:
http://www.usgbc.org/rpc

A LEED project team does not have to do anything special, since LEED v4 Online will automatically decide which RPC your project will get once you enter the project zip code and other information. If you have more than 4 RPCs, then you need to decide which 4 RPCs you want to use for your project.

The USGBC has been working on developing similar RPC incentives for international projects. RPC is available for many countries.

3. **AIA series of agreements for sustainable projects**
   AIA has developed a new series of agreements for sustainable projects:
   - A101- 2017 SP Standard Form of Agreement Between Owner and Contractor, Sustainable Project, Stipulated Sum Payment, Standard Form
   - A201 - 2017 SP General Conditions of the Contract for Construction, for use on a Sustainable Project
   - A401 - 2017 SP Standard Form of Agreement Between Contractor and Subcontractor, for use on a Sustainable Project
   - B101 - 2017 SP Standard Form of Agreement Between Owner and Architect, for use on a Sustainable Project
   - B214 - 2012 Standard Form of Architect' Services: LEED® Certification
   - C401- 2017 SP Standard Form of Agreement Between Architect and Consultant, for use on a Sustainable Project

   Many contracts require some or all of the design team members to be LEED certified. Architects should modify this requirement since not all team members need to be LEED accredited for the project to attain certification.

   Promise of a particular outcome such as the specific performance of a sustainable project, or guaranteeing a project will achieve a certain level of certification, can lead to litigations later. An architect should check the contract's language with care to avoid these pitfalls.

Chapter Four

Codes & Regulations (16-22%)

NCARB Requirements
In this section, you'll look at the codes and regulations appropriate to the initial analysis and programming phase of a project.

A. Identify relevant code requirements for building and site types (U/A)

NCARB Requirements
As an architect, you need to be able to determine and understand the codes and regulations that govern a specific project type and geographic region. This includes distinguishing relevant accessibility and ADA requirements, applicable building and energy codes, and sustainability related standards and guidelines. It's important to conduct an initial code analysis to understand how the codes will affect a project's design.

There is a very important concept in code development that states that existing buildings do not have to comply with new codes enacted after the buildings are built because they are "**grandfathered in**." The grandfather concept was created to prevent excessive hardship to existing building owners. This code is important, which ARE exams will test in one way or another.

Note:
When the term "code" or "codes" is used in this book, this term refers to all regulations, codes, laws, etc. issued by governing agencies or code publishers.

Even if a building is substantially remodeled, the cost of bringing the building into compliance with ADA should not be more than 20% to avoid undue hardship to the owner. If ARE exams test you on this concept, the 20% number will be buried somewhere in the materials provided in the exam. You can simply use the search function to find it.

1. Accessibility and ADA requirements

ADA also follows the "grandfathered in" rule of codes, *except* that it *does* require businesses that serve the public to remove barriers in existing facilities based on the following considerations.

Restaurant with a step at the entrance is not accessible to customers who use wheelchairs or scooters.

Restaurant after a ramp is installed, making the entrance accessible.

Figure 4.1 Making the entrance accessible (Source: ADA.gov)

Removing barriers
Removing or correcting barriers can be simple and inexpensive in one facility, but difficult and costly in another. For this reason, the ADA sets out a flexible rule for removing barriers. Businesses must remove physical barriers in existing facilities to improve accessibility where it is "readily achievable" to do so. Barrier removal is considered "readily achievable" when it can be easily accomplished, without much difficulty or expense.

A man who is blind is leaving a real estate agency. A workman is trimming a tree branch that hangs too low over the sidewalk.

Figure 4.2 Trimming a tree branch that hangs too low over the sidewalk (Source: ADA.gov)

"Readily achievable"
What is readily achievable is determined on a case-by-case basis, relative to a particular business's resources and existing barriers. Something readily achievable for one business may not be readily achievable for the one next door, because of more limited resources or more difficult physical constraints. And, what is readily achievable when a business is doing well may not be readily achievable in a down cycle when business is slow. A barrier that cannot be removed when business is slow should be reevaluated when business improves.

If a business has more barriers than it can afford to remove all at once, it can spread the work out over time. Removing barriers is an ongoing responsibility, so you should reevaluate the barriers in your facility every year to determine which ones to remove.

Priorities
The following is an example of a grocery store, but the concept of priority is applicable to all building types.

In evaluating what barriers need to be removed, you should give first priority to getting customers with disabilities in the door. The second priority is providing access to the areas where you provide goods and services to the public. The third priority is providing access to the restroom facilities that are provided for customers' use. Lastly, you should eliminate any other physical barriers you have.

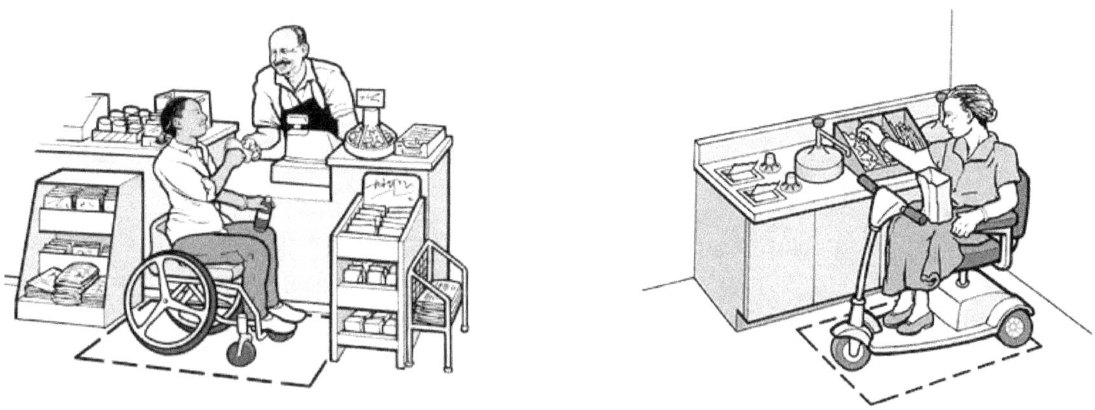

A woman using a wheelchair is buying a drink at a small grocery store. Adequate maneuvering space and a low, uncluttered counter make it possible for her to approach the sales counter.

A woman using a scooter is selecting items from the condiment bar at a casual restaurant. Adequate maneuvering space and items placed within reach make it possible for her to serve herself.

Figure 4.3 Providing adequate maneuvering spaces and a low counter (Source: ADA.gov)

Limits
The ADA does have limits for barrier removal. While stores may need to rearrange racks and shelves, restaurants may need to rearrange tables, and other businesses may need to rearrange furniture or potted plants in order to permit wheelchair access, businesses are not expected to reduce the amount of furniture or display racks to the extent that it results in a significant loss of selling or serving space.

In removing barriers, you never have to do more than is required under the standards for alterations.

The 20% undue hardship rule is also per ADA.Gov:

"When passed in 1990, the ADA did not require that all existing buildings be retrofitted for accessibility immediately. Instead, it relies on planned alterations and barrier removal to improve accessibility in older buildings over time. It says that when a business alters an existing facility in a way that affects usability, the areas or elements being altered must comply with the ADA Standards...

When a primary function area is altered, the path of travel to the altered area and the amenities serving the altered area must be made accessible, unless the costs for these changes are disproportionate. The costs for the added alterations are considered disproportionate if they exceed 20 percent of the cost of the overall alteration. In this case, you should make as many of the changes as you can without going over the 20 percent limit. Use this order of priority: entrance; route to the primary function area; at least one unisex restroom or one restroom for each sex serving the area; public telephones serving the area; drinking fountains serving the area; other elements."

See link:
https://www.ada.gov/reachingout/lesson33.htm

Accessibility and ADA requirements have **two main components**. The first is from a public way such as a public sidewalk, or a handicap parking stall to the main building entrance. The second is from the building entrance to the areas of various interior functions.

Accessibility and ADA requirements for this division are covered by the following codes:

- ***2010 ADA Standards for Accessible Design,*** U.S. Department of Justice, 2010
  Instead of repeating the content of the entire code, we'll discuss some of the most critical information for architectural practice and ARE exams.

  You can access this standard for free at the following link:
  https://www.ada.gov/2010ADAstandards_index.htm

  You can download the free 279-page PDF file for *2010 ADA Standards for Accessible Design*, and the free 170-page PDF file for *Guidance on the 2010 ADA Standards for Accessible Design*.

2010 ADA Standards for Accessible Design includes the following sections:

- Introduction
- 2010 Standards for State and Local Government Facilities: Title II
- 2010 Standards for Public Accommodations and Commercial Facilities: Title III
- 2010 Standards for Titles II and III Facilities: 2004 ADAAG

Note:

You do not need to read 2010 ADA Standards for Accessible Design line by line or from cover to cover. That will consume too much of your time, and you may NOT gain anything. Skim through it a few times, jump back and forth if you need to, and focus on the portions that you do not know or are not familiar with. Try to take as much ***critical*** *information as possible from the book* ***without*** *reading it line by line or from cover to cover. Focus on the* ***images, diagrams and charts****. If you already understand them, then move on to the next page. If you do not understand, then stop and read the associated text. This will help you cut out many pages of reading.*

You goal is NOT to read the standard line by line or from cover to cover, you goal is to prepare and pass this ARE division with your ***limited*** *time and effort.*

Think like an architect: pay attention to the stuff that you will need to know as an architect. You do not have to memorize everything. You just need to become familiar with this standard. Know where to find the information and how to use the information when needed.

The following are a few examples of diagrams from this standard: (All the diagrams in this standard are in public domain, and the U.S. government encourages you to copy and use them. You can feel free to copy and paste them onto your office's standard set of drawings as part of the standard project requirements.)

Advisory 303.3 Beveled. A change in level of ½ inch (13 mm) is permitted to be ¼ inch (6.4 mm) vertical plus ¼ inch (6.4 mm) beveled. However, in no case may the combined change in level exceed ½ inch (13 mm). Changes in level exceeding ½ inch (13 mm) must comply with 405 (Ramps) or 406 (Curb Ramps).

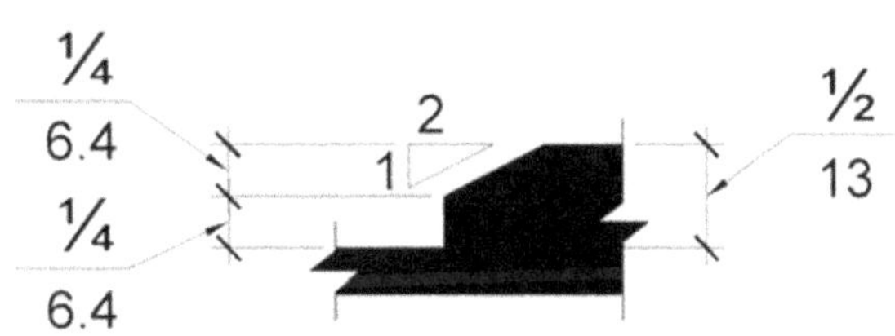

Figure 4.4 Beveled change in level advisory (Source: ADA.gov)

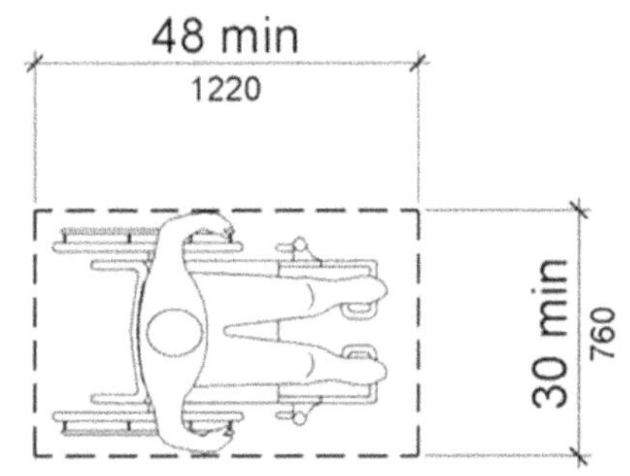

Figure 4.5 Clear floor or ground space (Source: ADA.gov)

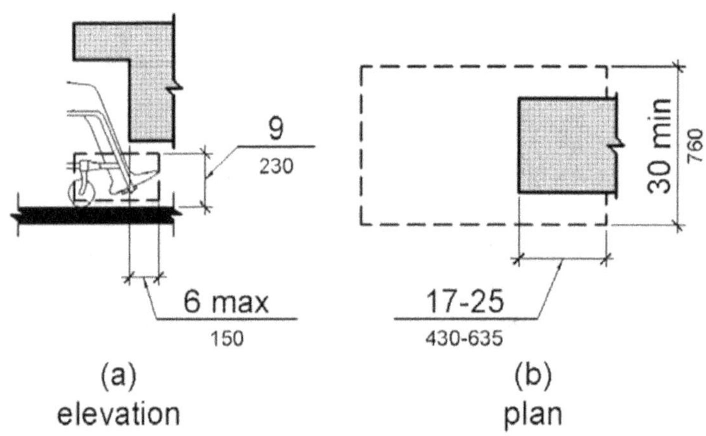

Figure 4.6 Toe clearance (Source: ADA.gov)

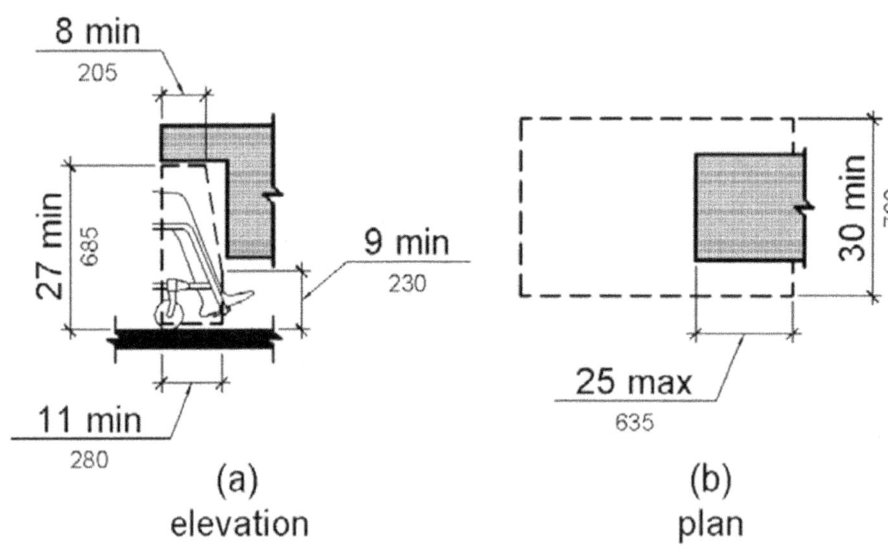

Figure 4.7 Knee clearance (Source: ADA.gov)

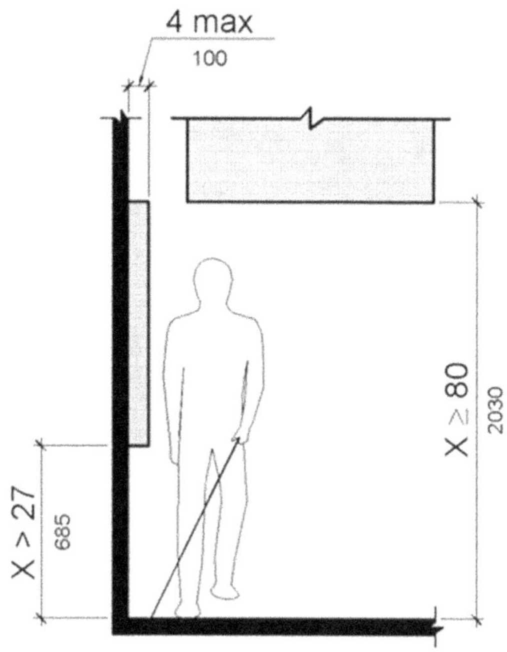

Figure 4.8 Limits of protruding objects (Source: ADA.gov)

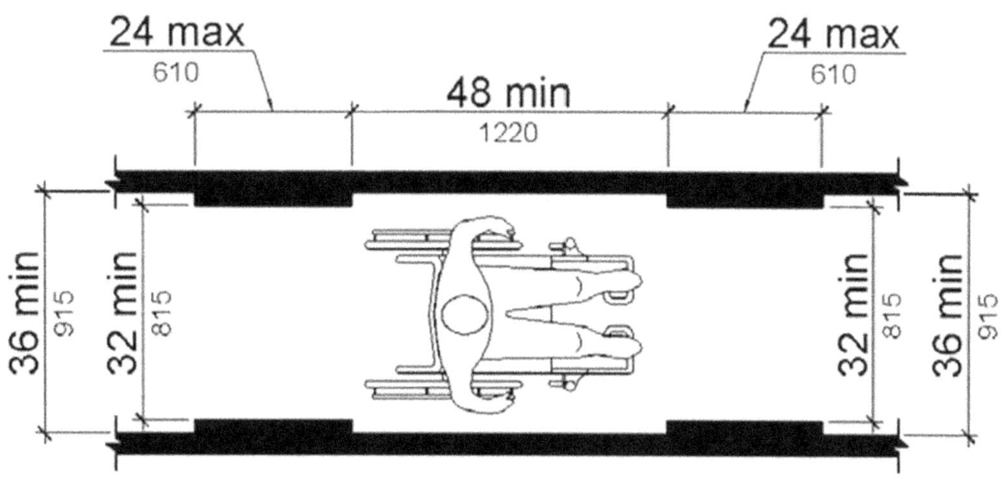

Figure 4.9 Clear width of an accessible route (Source: ADA.gov)

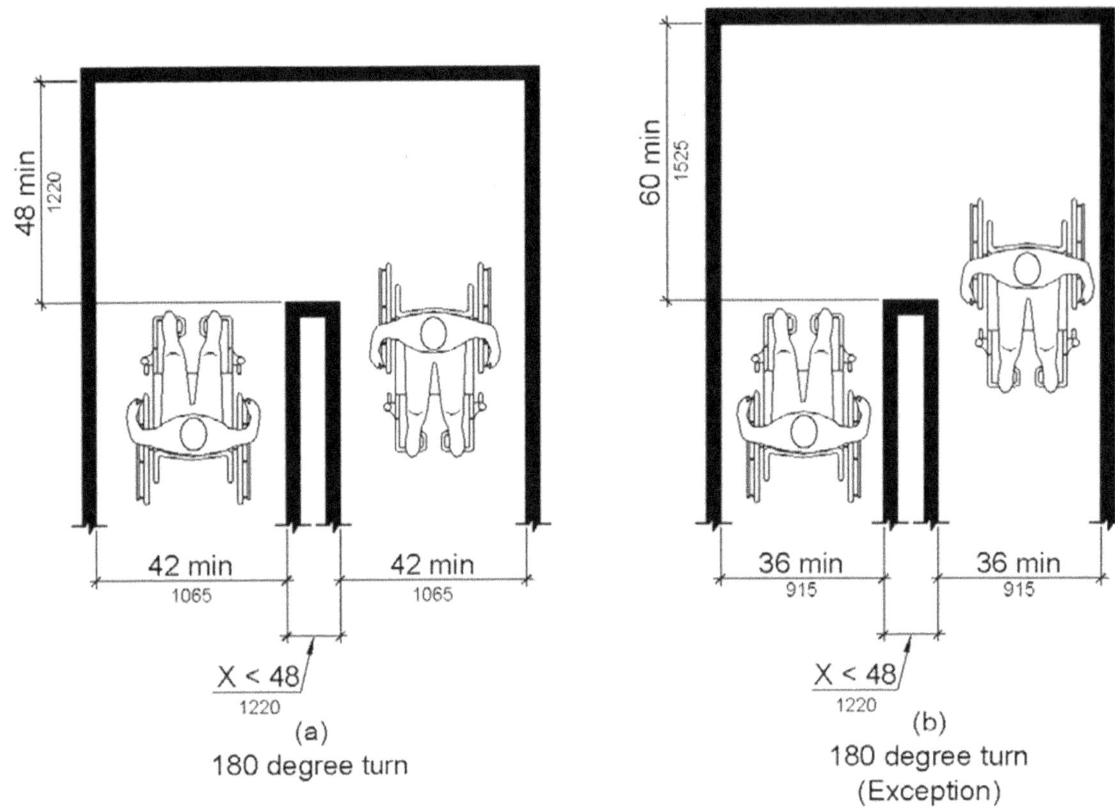

Figure 4.10 Clear width at turn (Source: ADA.gov)

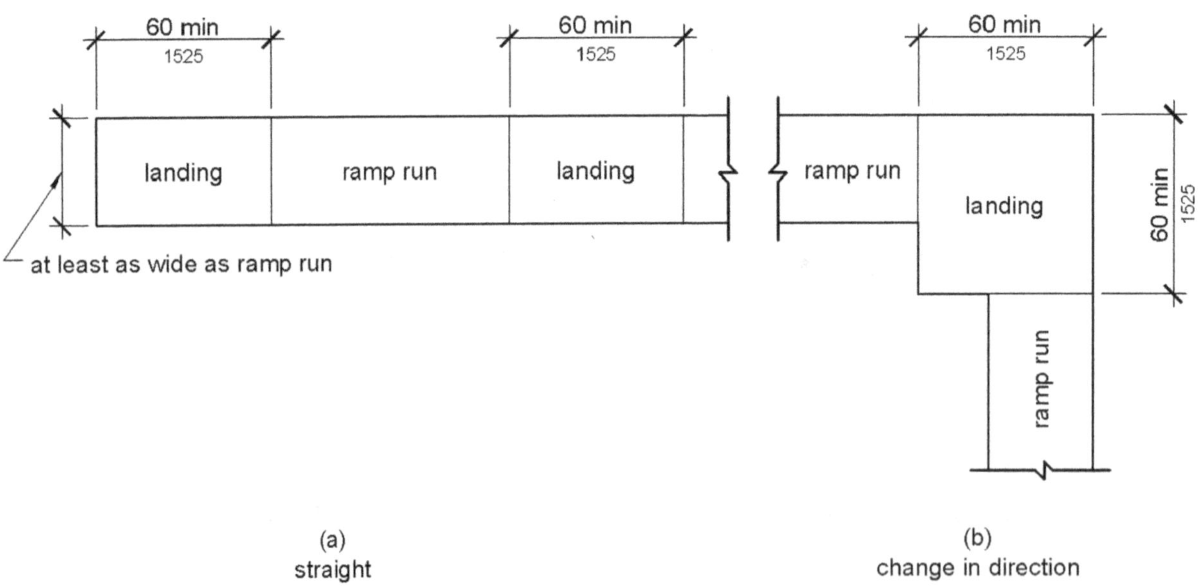

Figure 4.11 Ramp landings (Source: ADA.gov)

Advisory 405.9.1 Extended Floor or Ground Surface. The extended surface prevents wheelchair casters and crutch tips from slipping off the ramp surface.

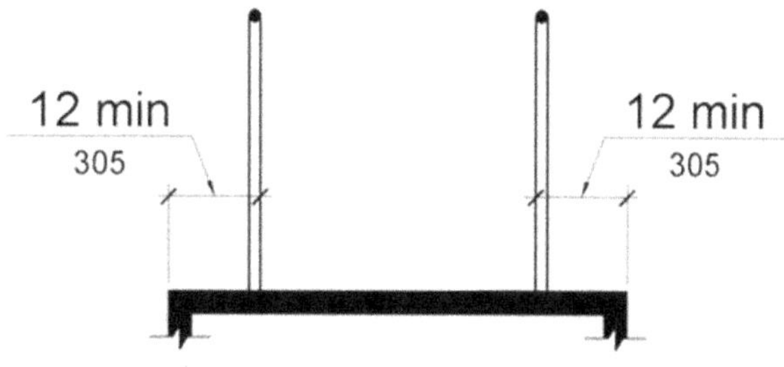

Figure 4.12 Extended floor or ground surface edge protection (Source: ADA.gov)

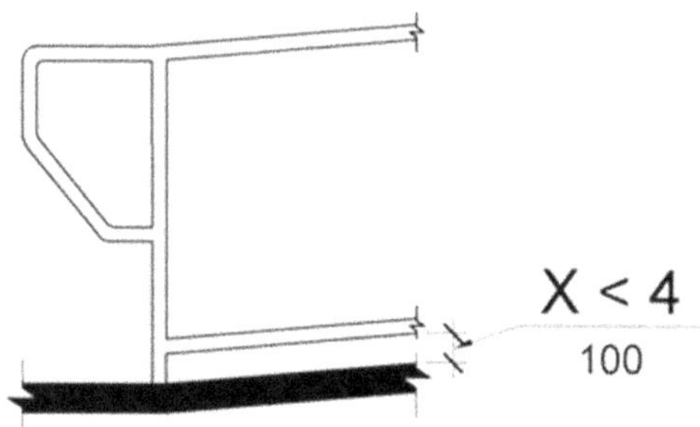

Figure 4.13 Curb or barrier edge protection (Source: ADA.gov)

502.2 Vehicle Spaces. Car parking *spaces* shall be 96 inches (2440 mm) wide minimum and van parking *spaces* shall be 132 inches (3350 mm) wide minimum, shall be marked to define the width, and shall have an adjacent access aisle complying with 502.3.

> **EXCEPTION:** Van parking *spaces* shall be permitted to be 96 inches (2440 mm) wide minimum where the access aisle is 96 inches (2440 mm) wide minimum.

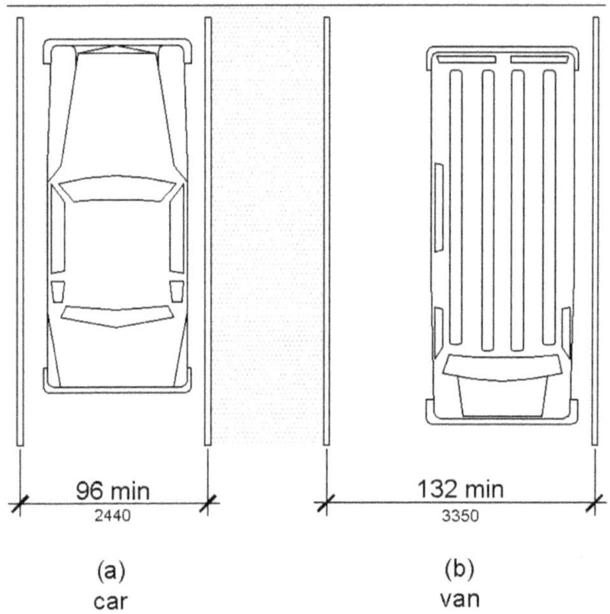

Figure 4.14 Vehicle parking spaces (Source: ADA.gov)

Advisory 502.3 Access Aisle. Accessible routes must connect parking spaces to accessible entrances. In parking facilities where the accessible route must cross vehicular traffic lanes, marked crossings enhance pedestrian safety, particularly for people using wheelchairs and other mobility aids. Where possible, it is preferable that the accessible route not pass behind parked vehicles.

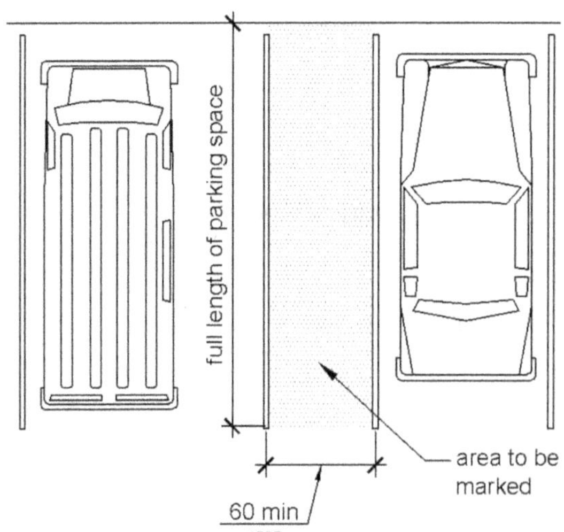

Figure 4.15 Parking space access aisle (Source: ADA.gov)

Total Number of Parking Spaces Provided in Parking Facility	Minimum Number of Required Accessible Parking Spaces
1 to 25	1
26 to 50	2
51 to 75	3
76 to 100	4
101 to 150	5
151 to 200	6
201 to 300	7
301 to 400	8
401 to 500	9
501 to 1000	2 percent of total
1001 and over	20, plus 1 for each 100, or fraction thereof, over 1000

Advisory 208.2 Minimum Number. The term "parking facility" is used Section 208.2 instead of the term "parking lot" so that it is clear that both parking lots and parking structures are required to comply with this section. The number of parking spaces required to be accessible is to be calculated separately for each parking facility; the required number is not to be based on the total number of parking spaces provided in all of the parking facilities provided on the site.

Figure 4.16 Parking spaces (Source: ADA.gov)

Guidance on the 2010 ADA Standards for Accessible Design includes the following sections titled:

- State and Local Government Facilities: Guidance on the Revisions to 28 CFR 35.151
- Public Accommodations and Commercial Facilities: Guidance on the Revisions to 28 CFR part 36, subpart D
- Analysis and Commentary on the 2010 ADA Standards for Accessible Design
- Section-by-Section Analysis with Public Comment

Figure 4.17 and figure 4.18 are examples of the diagrams from *Guidance on the 2010 ADA Standards for Accessible Design.*

- ***ICC A117.1-2009 Accessible and Usable Buildings and Facilities,*** International Code Council, 2010

  You can access the code for free at this link: (This code is protected by copyright, so you can review the information but cannot copy or download it without prior written permission from ICC.)
  https://codes.iccsafe.org/content/icca117-12009

ICC A117.1-2009 Accessible and Usable Buildings and Facilities has a lot of overlap with *Accessibility and ADA requirements.* You should look through it two or three times. Once again, only focus on the diagrams and tables.

This drawing shows an accessible 13-foot wide guest room with features that comply with the 2010 Standards. Features include a standard bathtub with a seat, comparable vanity, clothes closet with swinging doors, and door connecting to adjacent guest room. Furnishings include a king bed and additional seating.

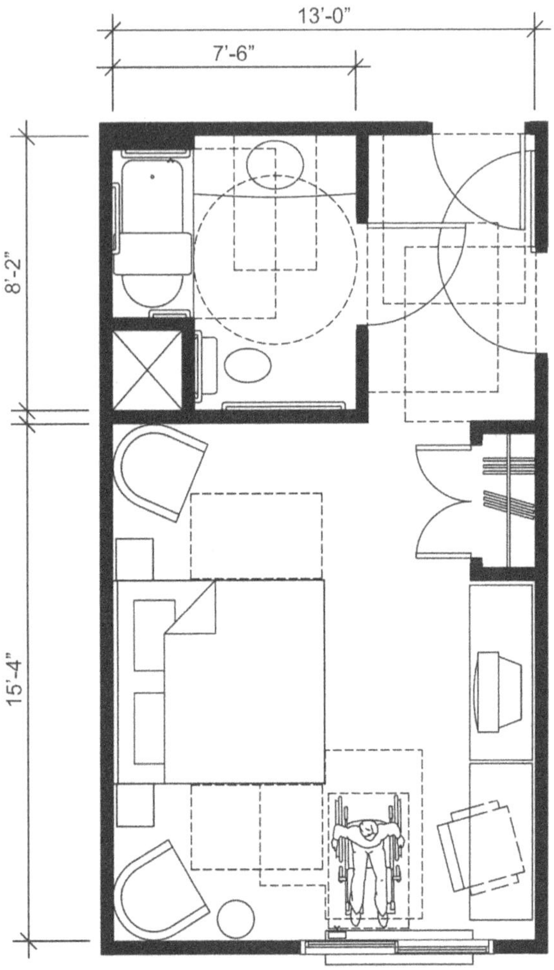

The following accessible features are provided in the bathroom:

- Comparable vanity counter top space (section 806);
- Bathtub with a lavatory at the control end (section 607.2);
- Removable bathtub seat (section 607.3);
- Clearance in front of the bathtub extends its full length and is 30 inches wide min. (section 607.2);
- Recessed bathtub location permits shorter rear grab bar at water closet (section 604.5.2);
- Circular turning space in room (section 603.2.1);
- Required clear floor spaces at fixtures and turning space overlap (section 603.2.2);
- Turning space includes knee and toe clearance at lavatory (section 304.3);
- Water closet clearance is 60 inches at back wall and 56 inches deep (section 604.3);
- Centerline of the water closet at 16-18 inches from side wall (section 604.2); and
- No other fixtures or obstructions located within required water closet clearance (section 604.3).

The following accessible features are provided in the living area:

- T-shaped turning space (section 304.3.2);
- Accessible route (section 402);
- Clear floor space on both sides of the bed (section 806.2.3);
- Maneuvering clearances at all doors (section 404.2);
- Accessible operable window (section 309); and
- Accessible controls for the heat and air conditioning (section 309).

Figure 4.17 Plan 1A: thirteen-foot wide accessible guest room (Source: ADA.gov)

This drawing shows an accessible 13-foot wide guest room with features that comply with the 2010 Standards. Features include a standard bathtub with a seat, comparable vanity, clothes closet with swinging doors, and door connecting to adjacent guest room. Furnishings include two beds.

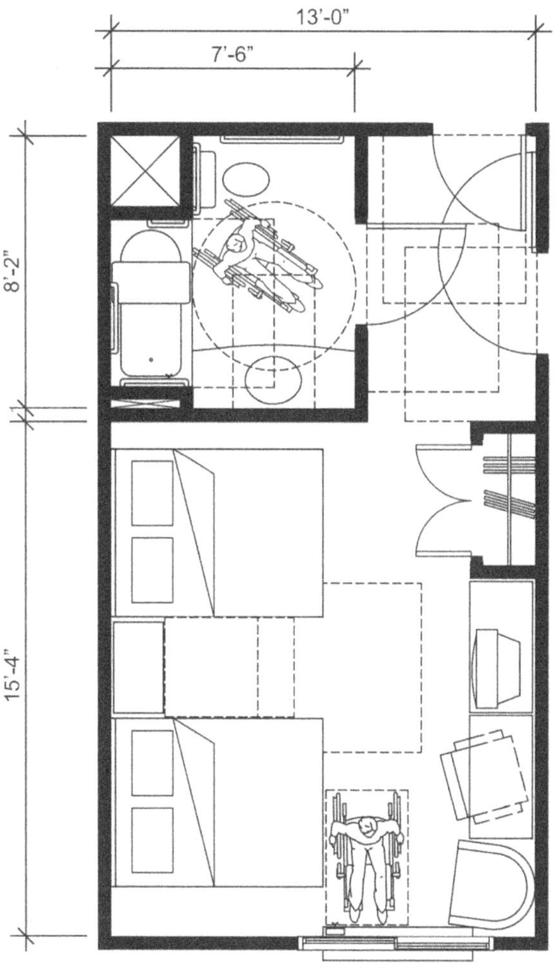

The following accessible features are provided in the bathroom:

- Comparable vanity counter top space (section 806);
- Bathtub with a lavatory at the control end (section 607.2);
- Removable bathtub seat (section 607.3);
- Clearance in front of the bathtub extends its full length and is 30 inches wide min. (section 607.2);
- Recessed bathtub location permits shorter rear grab bar at water closet (section 604.5.2);
- Circular turning space in room (section 603.2.1);
- Required clear floor spaces at fixtures and turning space overlap (section 603.2.2);
- Turning space includes knee and toe clearance at lavatory (section 304.3);
- Water closet clearance is 60 inches at back wall and 56 inches deep (section 604.3);
- Centerline of the water closet at 16-18 inches from side wall (section 604.2); and
- No other fixtures or obstructions located within required water closet clearance (section 604.3);

The following accessible features are provided in the living area:

- T-shaped turning space (section 304.3.2);
- Accessible route (section 402);
- Clear floor space between beds (section 806.2.3);
- Maneuvering clearances at all doors (section 404.2);
- Accessible operable window (section 309); and
- Accessible controls for the heat and air conditioning (section 309).

Figure 4.18 Plan 1B: thirteen-foot wide accessible guest room (Source: ADA.gov)

2. **IBC chapters and sections that you should pay special attention to**
Per the latest NCARB handbook, the IBC in use for the current ARE exams is *International Building Code (2018)*, published by International Code Council, 2017.
You can access the current code for free at this link:
https://codes.iccsafe.org/content/IBC2018

You should read the section entitled "Effective Use of the International Building Code" carefully. It tells you which chapter of the IBC is correlated to another code such as IFC, IMC, and IPC. This is available at the following link:
https://codes.iccsafe.org/content/IBC2018/effective-use-of-the-international-building-code

Every project will require code analysis, and this topic is important content for the PA exam. Code analysis is typically drawn or notated on the cover sheet of a project.

Code analysis is difficult for young professionals. Code books have hundreds or even thousands of pages. It is hard to know where to begin.

If you have done it once or twice, it will be much easier for you.

Why?

Because there are only a few code sections that are frequently used. If you have done it once, you will know where to find the information and how to do it.

We are going to use a sample code analysis for a retail building to help you to become familiar with this process. You may be doing other types of building, but there are many universal requirements for all buildings that you can learn from this sample. The codes may be updated in the future, and the code section numbers may change, but if you can learn from this sample and know what you are looking for, you can easily locate the information from the **index** of the new version of the code books in the future.

The following is a sample **code analysis** for a retail building:

Applicable Building Codes:

2018 International Building Code (IBC)

2018 National Electrical Code (NEC)

2018 International Mechanical Code (IMC)

2018 International Plumbing Code (IPC)

Energy Code: California Title 24 Energy

Accessibility Code: California Title 24 Handicap

Building Zoning: Commercial

Occupancy Group: M (Section 309, Mercantile Group M, 2018 IBC)

Number of Stories: 1

Type of Construction: V-B (Table 601, Fire-Resistance Rating Requirements for Building Elements; Table 602, Fire-Resistance Rating Requirements for Exterior Walls Based on Separation Distance, 2018 IBC), Fully Sprinklered

Building Height: Main Building 22'-0" < 40'-0"
Canopy 26'-0" < 40'-0"
(Table 504.3, Allowable Building Height in Feet above Grade Plane, 2018 IBC)

Qualification of Building:

Allowable Area: 36,000 s.f. (Table 506.2, Allowable Area Factor in Square Footage, 2018 IBC. Occupancy Group: M, Fully Sprinklered and 1 story in height: S1, Type of Construction:
V-B)

Actual Building Area = 8,806 s.f. < 36,000 s.f.

If you are doing the building shell only, the code analysis above is probably adequate. If you are also doing tenant improvement work for the building, you probably need to do the following additional code analysis:

Common Path of Egress Travel Distance:

Maximum Travel Distance Allowed: 250 ft. for fully sprinklered, M occupancy building (Table 1017.2, Exit Access Travel Distance, 2018 IBC)

Maximum Travel Distance Provided: 73 ft. < 250 ft.

Note:
You may need to draw a dashed polyline on your floor plan to show the worst-case scenario of your occupants' Common Path of Egress. You can measure the total length of the dashed polyline to get the actual travel distance. Common path of egress is rarely a straight line because you have fixtures or shelving blocking the direct travel path in most cases. So, you probably have to zigzag your polyline. We have to do this per plan checker's comments many times in our daily

practices. I'd not be surprised to see some real ARE exam questions on this section.

Occupant Load
(Table 1004.1.2, Maximum Floor Area Allowances per Occupant, 2018 IBC):

Function of Space	Area (s.f.)	Occupant Load Factor	Number of Occupants
Mercantile	5,575	60 gross	5,575/60 = 93
Receiving	1,893	300 gross	1,893/300 = 7
Office (Business)	1,338	100 gross	1,338/100 = 14
Total Occupant Load			114 (Male: 114/2 = 57, Female: 114/2 = 57)

Figure 4.19 Occupant load calculations

Note:
You can use AutoCAD or Revit to easily calculate the square footage of each kind of space for your project.

Minimum Number of Exits

Number of Exits Required: 2 for occupant load between 49 and 501 for M occupancy.
(Table 1006.2.1, Spaces of One Exit or Exit Access Doorway, 2018 IBC)

Note:
Three exits or exit access doorways shall be provided from any space with an occupant load of 501 to 1,000. Four exits or exit access doorways shall be provided from any space with an occupant load greater than 1,000.

M occupancy group allows one exit if the occupant load is 49 or less. Since we have 114 occupants and the occupancy is larger than 49 but smaller than 501, we need two exits.

Number of Exits Provided: 3 > 2 (Check your project and fill in the actual number of exits provided here.)

Egress Width

All exit doors shall have panic hardware.

Egress Width Required: 114 x 0.2 = 22.8" (or 1'-11")
(Section 1005.3.2, Other egress components, 2018 IBC)

Egress Width Provided: 3' + 3' + 6' = 12' > 1'-11"

Minimum Number of Plumbing Fixtures
(Table 2902.1, Minimum number of required plumbing fixtures, 2018 IBC)

(See previous occupant load calculation)

Toilets (For M Occupancy, 1 per 500 occupants per sex)

Required:	Male	1 (57 occupants < 500)	Female	1
Provided:	Male	1	Female	1

Lavatories (For M Occupancy, 1 per 750 occupants per sex)

Required:	Male	1 (57 occupants < 750)	Female	1
Provided:	Male	1	Female	1

Drinking Fountains (For M Occupancy, 1 per 1,000 occupants)
Required: 1 (114 occupants for both sexes < 1,000)
Provided: 1

Other (For M Occupancy, 1 service sink required, 1 service sink provided)

Several areas of the IBC that you can focus on are:

- **Chapter 3, Use and occupancy classification**
  You need to understand the definition of each occupancy group and know how to select the correct occupancy group for a building.

- **Chapters 5 and 6, General building heights and areas and types of construction**
  You need to understand and know how to use the tables and sections. Read them a few times to become familiar and know where to find the information. You do not need to memorize them.
  - Table 504.3, Allowable Building Height in Feet above Grade Plane
  - Table 504.4, Allowable Number of Stories above Grade Plane
  - Table 506.2, Allowable Area Factor in Square Feet
  - Section 507, Unlimited Area Building

  Note:
  You need to know how yards are measured. For many occupancy groups, if a one-story building is surrounded and adjoined by yards or public ways not less than 60 ft. in width, it can have an unlimited area. This section is extremely useful in real practice and probably tested in ARE exams.

  For example, when I was working as a senior project manager in a major retail firm, we did many retail buildings. We always qualified the entire shopping center as "one building" with 60 ft. wide yards (public ways) around all sides, even though a shopping center typically has many buildings within it. This is the cheapest way to build because you can use Type V construction for every building in the shopping center. I think most, if not all, of the shopping centers use this IBC section to qualify, at least on the West Coast.

- Table 508.4, Required Separation of Occupancy
- Table 601, Fire-Resistance Rating Requirements for Building Elements
- Table 602, Fire-Resistance Rating Requirements for Exterior Walls Based on Separation Distance

Note:
Pay attention to the notes and exceptions for each table.

- **Chapters 7 to 9, Fire and smoke protection features, Interior finishes, and Fire protection systems**
  Areas to become familiar with are as follows:
  - Table 705.2, Minimum Distance of Projection
  - Table 705.8, Maximum Areas of Exterior Openings Based on Fire Separation Distance and Degree of Opening Projection
  - Section 718, Concealed Spaces: pay attention to Section 718.2, fire blocking, Section 718.3, draftstopping in floors, and Section 718.4, draftstopping in attics.
  - Table 721.1, Rate Fire-Resistance Periods for Various Walls and Partitions
  - Table 803.11, Interior Wall and Ceiling Finish Requirements by Occupancy

- **Chapter 10, Means of egress**
  - Table 1004.1.2, Maximum Floor Area Allowances per Occupant

    Note:
    This table is very important. You can use it to calculate the number of occupants for a space (the occupant load), and then in turn calculate the exit width and number of plumbing fixtures required.

  - Section 1005.3.1, Stairways: The capacity, in inches, of means of egress stairway shall be calculated by the occupant load served by such stairways by a mean of egress capacity factor of 0.3 inch per occupant. When stairway serve more than one story, only the occupant load of each story considered individually shall be used in calculating the required capacity of the stairway serving that story.

    Note:
    You need to pay attention to the exceptions.

  - Section 1005.3.2, Other egress components. The capacity, in inches, of means of egress components *other than* stairways shall be calculated by the occupant load served by such components by a mean of egress capacity factor of 0.2 inch per occupant.

    Note:
    You need to pay attention to the exceptions.

  - Table 1006.2.1, Spaces of One Exit or Exit Access Doorway.

- o Table 1006.3.1, Minimum Number of Exits or Access to Exits per Story
- o Section 1007.1.1, Two Exits or Exit Access Doorways

Note:
The distance between the two exits shall be more than half of the maximum diagonal dimension of the building or area to be served. You need to pay attention to the exceptions.

- o Table 1017.2, Exit Access Travel Distance
- o Table 1020.1, Corridor Fire-Resistance Rating
- o Table 1020.2, Minimum Corridor Width
- o Section 1020.4, Dead ends: Dead end corridors shall *not* be more than 20 ft. in length.
- o Section 1024, Exit Passageways
- o Section 1026, Horizontal Exits
- o Section 1028, Exit Discharge
- o Section 1029.7, Travel distance: 200 ft. maximum measured along line of travel in non-sprinklered buildings, and 250 ft. maximum measured along line of travel in sprinklered buildings.

- **Chapter 11, Accessibility (specific requirements to allow use and access to a building for persons with disabilities)**
  There is a lot of overlap between the IBC, 2010 ADA Standards for Accessible Design, and ICC A117.1-2009 Accessible and Usable Buildings and Facilities regarding accessibility requirements. Yet they also have some differences. You need to comply with the most restrictive requirement.

  Become familiar with the following:
  - o Table 1106.1, Accessible Parking Spaces

- **Chapter 12, Interior Environment**
  - o Section 1203, Ventilation: Pay attention to Section 1203.2 (net free ventilation area shall be not less than 1/150 of the area of the spaces vented), and Section 1203.5.1 (the openable areas of the opening to the outdoor shall not be less than 4 percent of the floor areas being vented).

- **Chapter 29, Plumbing systems (fixtures)**
  - o Table 2902.1, Minimum Number of Required Plumbing Fixtures

  Note:
  This table is often used with Table 1004.1.2, Maximum Floor Area Allowances per Occupant, to determine the minimum number of plumbing fixtures needed.

3. **IRC**
   International Residential Code (IRC), International Green Construction Code (IGCC), and energy codes are mentioned in ARE handbook, PA Section 2, Codes and Regulations, but not listed in PA References, so we'll briefly discuss them. You just need to know where to find them, and when and how to use them.

   You can find the latest free version of IRC at the following link:
   https://codes.iccsafe.org/content/IRC2018P3

   When adapted by a local jurisdiction, IRC establishes minimum standards for one- and two-family dwellings and townhouses using prescriptive provisions.

4. **IGCC**
   You can find the latest free version of IGCC at the following link:
   https://codes.iccsafe.org/content/IGCC2018P2

   IGCC establishes minimum standards for building systems and site consideration using prescriptive and performance-related provisions.

5. **Energy codes**
   You can find the latest free version of International Energy Conservation Code (IECC) at the following link:
   https://codes.iccsafe.org/content/IECC2018P3

   IECC has separate provisions for commercial buildings and low-rise residential buildings (three floors or less in height above grade).

   You can find the latest free version of International Solar Energy Provisions (ISEP) at the following link:
   https://codes.iccsafe.org/content/ISEP2018

   ISEP brings together all solar energy provisions for both solar thermal and photovoltaic systems.

   Several states such California, Ohio, Minnesota, and Florida also have their own energy codes.

   You can find the latest free version of them at the following link with a title key word search of "energy code":
   https://codes.iccsafe.org

   The latest free version of California Energy Code, Title 24, Part 6 is available at the following link:
   https://codes.iccsafe.org/content/CEC2019P1

B. Identify relevant zoning and land use requirements (U/A)

NCARB Requirements
You need to be able to determine and understand requirements that limit the extent of site and building development. This includes issues like setbacks, footprint limitations, maximum building heights, FAR, parking requirements, easements, and other zoning and land use regulations. Recognizing situations where a special exception or variance is suitable and understanding the process for approval are also included in this objective.

In the U.S., most states and cities adapted the ***International Building Codes*** **(IBC)**, but they also have their own special requirements.

At the beginning of the project, you need to check the city's website and call the various departments of the cities to do a quick code research. The codes that affect architectural design are typically part of the city's **municipal codes**, like parking requirements, setback requirements, landscape requirements, etc. You can simply download a copy of the **municipal codes** and read it. You can also call the various departments of the cities if you need help locating the information or if you have additional questions.

You can obtain the zoning and land use requirements from the planning department based on the address of your property.

Zoning is the process of dividing land of a city into zones (e.g. residential, commercial, mixed-use, industrial, institutional, recreational, open space) in which certain land uses are prohibited or permitted.

The various approaches to zoning can be divided into four main categories:

- **Euclidean zoning or single-use zoning** is named for a type of zoning code adopted in Euclid, Ohio, and approved in a landmark decision of the U.S. Supreme Court, Euclid v. Ambler. Euclidean zoning codes are the most prevalent in the United States. They are characterized by segregating land uses into specified geographic districts and dimensional standards designating limitations on development activity within each type of district. Advantages include ease of implementation, relative effectiveness, long-established legal precedent, and familiarity. However, Euclidean zoning has received criticism for institutionalization of now-outdated planning theory and lack of flexibility.
- **Performance zoning or "effects-based planning"** uses goal-oriented or performance-based criteria to establish planning review parameters. Performance zoning is intended to provide rationality, flexibility, transparency, and accountability. This type of zoning avoids the arbitrariness of Euclidean zoning and better accommodates private property rights and market principles with environmental protection, but requires a high level of discretionary activity on the part of the governing agencies. Performance zoning has not been widely adopted in the United States.
- **Incentive zoning** was first implemented in New York City and Chicago, and intended to provide a reward-based system to encourage development that met established urban development goals. Typically, this type of zoning establishes a base level of limitations and a reward scale to encourage developers to incorporate the desired amenities. The

governing agencies may allow a developer to build a larger footprint or more stories to gain more usable space (the incentive) if the developer builds some amenities that the community or the public can use, such as a public park. Incentive zoning can be complex to administer, but allows a high degree of flexibility.
- **Form-based zoning** offers considerably more governmental latitude in building form and use as opposed to Euclidean zoning. Form-based zoning regulates the form that land use may take instead of the type. For instance, form-based zoning in a dense area may insist on low density, setbacks, and pedestrian accessibility. Form-based zoning is designed to directly respond to the physical structure of a community to create a more adaptable and walkable environment.

Urban zones fall into one of five main categories:
- residential
- mixed residential and commercial
- commercial
- industrial
- spatial (e. g. power plants, sports complexes, airports, shopping malls etc.)

Each category can have a number of sub-categories.

Most cities have both zoning ordinances and zoning maps. For example, figure 4.20 is a city of Irvine zoning map and figure 4.21 is a partial city of Irvine zoning map. You can see the legend that shows some zoning categories and sub-categories. Zoning maps are often color-coded. A full-size color copy is available for download at the city's website:
https://www.cityofirvine.org/community-development/zoning

You can download the zoning codes and zoning map for the city where your project is located. However, it is always a good idea to confirm the latest requirements with a planner at the city.

The following are some common zoning requirements:

1. **Setbacks**
   The Planning Department may have different setback requirements for front yard, side yards, and rear yard. For example, a residential property may have a 20 ft. front yard setback, 5 ft. side yard setbacks, and 10 ft. rear yard setback.

2. **Footprint limitations**
   Site coverage or **lot coverage** is the size of the footprint(s) of a building(s) and/or structure(s) on a site divided by the area of the site and is expressed as a percentage. Site coverage is used in calculating the intensity of use of a parcel for a development project. For example, a building footprint of 1,000 sf. on a 5,000-sf. plat has a 20% site coverage.

   You need to know the allowed site coverage requirement or the maximum percentage of site area that can be covered by your building. A single-family home commonly has a site coverage of 50%.

3. **Maximum building heights**
Based on the project address and the related zoning, the planning department can tell you the maximum allowable building height. For example, some cities require a single-family home to have a maximum building height of thirty-five feet from grade or be less than three stories tall.

4. **FAR**
FAR or Floor Area Ratio is the total floor area of the buildings divided by the site area. Your city often has a maximum FAR allowed for your site.

5. **Parking requirements**
Ask the planning department the following questions:
   - What is the standard parking stall size? (8' x 18' or 9' x 19' is common)
   - What is the compact parking stall size? (7' x 15' or 8' x 16' is common)
   - What is the handicap parking stall size and the van's parking stall size?
   - What is the percentage of compact parking stalls allowed? Generally, 20% or 25% is common.
   - Will car front **overhang** be allowed?

   You can determine the ratio of accessible parking required based on the related accessibility standards the city has adopted.

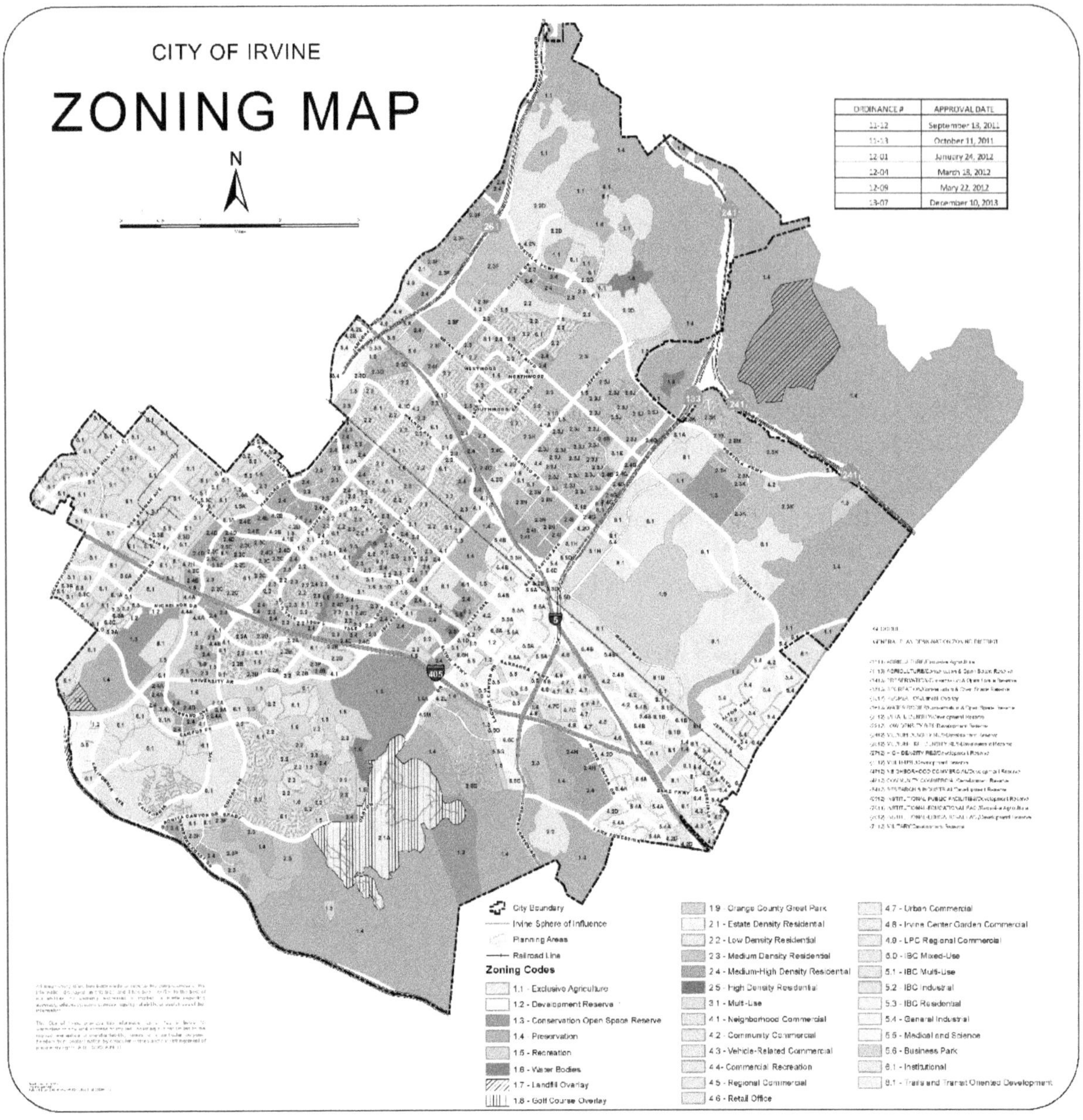

Figure 4.20 City of Irvine zoning map

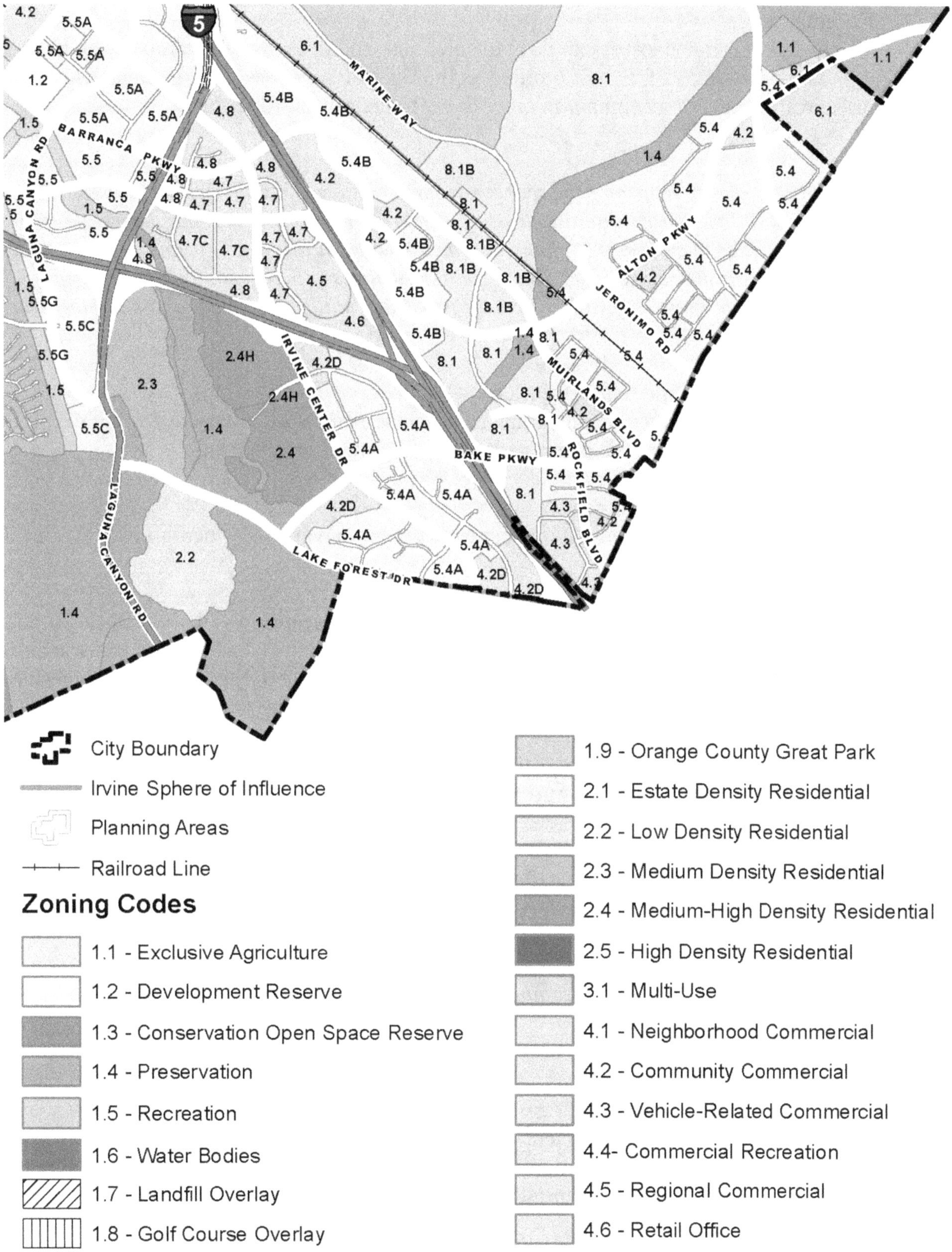

Figure 4.21 Partial city of Irvine zoning map

6. **Easements**

An **easement** is a nonpossessory right to enter onto and/or use the real property of another without possessing it. It is best typified as the right of way in which one landowner may pass over the land of a neighbor in order to get to their own property.

Note

Check your survey map or title report or check with your civil engineer or the utility companies to find out if your site has any ***easement requirements****, and determine if you can build within the easement. Some easements will not allow you to build at all, while other easements may allow you to build small structures with a maximum footing depth, or will allow you to build sidewalks or curbs or parking stalls. You need to find out the exact requirements.*

7. **Other zoning and land use regulations**

Other zoning and land use regulations may include some special types of zoning in rural or suburban areas, or special open space reservation requirements, etc.

8. **Special exception or variance**

The governing agencies may allow some exception or variance when a property owner shows an undue hardship.

The main difference between a **variance** and a **special exception** is that a variance grants a property owner the ability to use her or his property in a way that is completely against local regulations, while a special exception is a situation that governing agencies specifically recognize before creating a law, and will make provisions that recognize exceptions in the regulation itself. A special exception is a way to grandfather in an existing condition, like some old buildings that do not meet the current earthquake standards but were built before the standards were enacted.

C. Identify relevant local and site-specific requirements (U/A)

NCARB Requirements

You will need to determine and understand specialty regulations that apply to a particular site. These may include requirements about wetlands, flood plains, design overlay districts, or historic preservation.

1. **Wetlands**

Per EPA, wetlands are areas that are inundated or saturated by surface or ground water at a frequency and duration sufficient to support, and that under normal circumstances do support, a prevalence of vegetation typically adapted for life in saturated soil conditions. Wetlands generally include swamps, marshes, bogs, and similar areas.

The EPA and the Corps use the 1987 Corps of Engineers Wetlands Delineation Manual and Regional Supplements to define wetlands for the **Clean Water Act Section 404** (CWA 404) permit program (figure 4.22). Section 404 requires a permit from the Corps or authorized state for the discharge of dredged or fill material into the waters of the United States, including wetlands.

The 1987 Corps of Engineers Wetlands Delineation Manual and Regional Supplements organizes characteristics of a potential wetland into three categories: soils, vegetation and hydrology. The manual and supplements contain criteria for each category. With this approach, an area that meets all three criteria is considered a wetland.

Jurisdictional Determinations are issued by the Army Corps of Engineers, and determine whether a water will be regulated under CWA 404. These are often determined by performing a jurisdictional delineation of waters on a property.

You can find EPA wetland regulations and related information at the following links:
https://www.epa.gov/wetlands/wetlands-regulations
https://www.epa.gov/cwa-404/how-wetlands-are-defined-and-identified-under-cwa-section-404

Per the U.S. Geological Survey (**USGS**), the people of the United States have begun to recognize that wetlands have numerous and widespread benefits. However, many of the goods and services wetlands provide have little or no market value. Because of this, the benefits produced by wetlands accrue primarily to the general public. Therefore, the Government provides incentives and regulates and manages wetland resources to protect the resources from degradation and destruction. Other mechanisms for wetland protection include acquisition, planning, mitigation, disincentives for conversion of wetlands to other land uses, technical assistance, education, and research.

Although many States have their own wetland regulations, the Federal Government bears a major responsibility for regulating wetlands. The five Federal agencies that share the primary responsibility for protecting wetlands include the Department of Defense, U.S. Army Corps of Engineers (Corps); the U.S. Environmental Protection Agency (EPA); the Department of the Interior, U.S. Fish and Wildlife Service (FWS); the Department of Commerce, National Oceanic and Atmospheric Administration (NOAA); and the Department of Agriculture, Natural Resources Conservation Service (NRCS) (formerly the Soil Conservation Service). Each of these agencies has a different mission that is reflected in the implementation of the agency's authority for wetland protection. The Corps' duties are related to navigation and water supply. The EPA's authorities are related to protecting wetlands primarily for their contributions to the chemical, physical, and biological integrity of the Nation's waters. The FWS's authorities are related to managing fish and wildlife-game species and threatened and endangered species. Wetland authority of NOAA lies in its charge to manage the Nation's coastal resources. The NRCS focuses on wetlands affected by agricultural activities.

The U.S. Geological Survey website has some useful information on wetlands at:
https://water.usgs.gov/nwsum/WSP2425/legislation.html

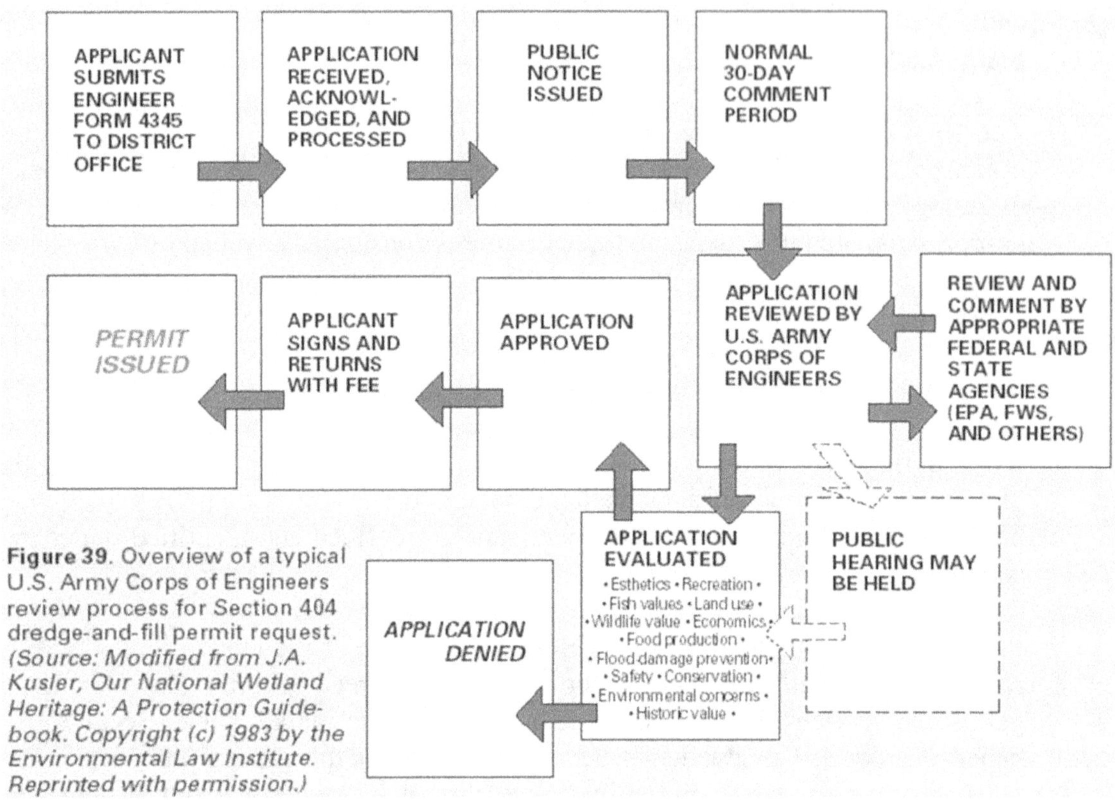

Figure 4.22 Overview of Corps of Engineers Wetlands permit process (Source: USGS)

2. **Floodplains**
   See detailed discussions in Chapter Three.

3. **Historic preservation**
   The Secretary of the Interior's Standards offer four distinct approaches to the treatment of historic properties with accompanying guidelines for each:
   - **Preservation:** Preservation is defined as the act or process of applying measures necessary to sustain the existing form, integrity, and materials of an historic property. Work, including preliminary measures to protect and stabilize the property, generally focuses upon the ongoing maintenance and repair of historic materials and features rather than extensive replacement and new construction. New exterior additions are not within the scope of this treatment; however, the limited and sensitive upgrading of mechanical, electrical, and plumbing systems and other code-required work to make properties functional is appropriate within a preservation project.
   - **Rehabilitation:** Rehabilitation is defined as the act or process of making possible a compatible use for a property through repair, alterations, and additions while preserving those portions or features which convey its historical, cultural, or architectural values.
   - **Restoration:** Restoration is defined as the act or process of accurately depicting the form, features, and character of a property as it appeared at a particular period of time by means of the removal of features from other periods in its history and reconstruction of missing features from the restoration period. The limited and sensitive upgrading of

mechanical, electrical, and plumbing systems and other code-required work to make properties functional is appropriate within a restoration project.

- **Reconstruction:** Reconstruction is defined as the act or process of depicting, by means of new construction, the form, features, and detailing of a non-surviving site, landscape, building, structure, or object for the purpose of replicating its appearance at a specific period of time and in its historic location. (Source: https://www.nps.gov)

See the Secretary of the Interior's Standards for Preservation, Rehabilitation, Restoration, and Reconstruction at the following link:
https://www.nps.gov/tps/standards.htm

Chapter Five

Site Analysis & Programming (21-27%)

NCARB Requirements
In this section, you'll need to analyze a project site relative to the program and project requirements.

A. Evaluate relevant qualitative and quantitative attributes of a site as they relate to a program (A/E)

NCARB Requirements
This objective assesses your ability to analyze a project site relative to the requirements in the program to determine if it is appropriate and feasible for development. You will need to consider the climate, topography, drainage, soil, built and natural features, utilities, access points, traffic patterns, easements, and other attributes relevant to the project and program.

1. Climate

Per National Weather Service, the earth's tilt, rotation and land/sea distribution affect the global weather patterns we observe. While the weather varies from day-to-day at any particular location, over the years, the same type of weather will reoccur. The reoccurring "average weather" found in any particular place is called climate.

German climatologist and amateur botanist Wladimir Köppen (1846-1940) divided the world's climates into categories based upon general temperature profile related to latitude. He worked with Rudolf Geiger to modify these categories which is known today as the Köppen-Geiger climate classification system. The major categories are as follows:

A - Tropical Climates
Tropical moist climates extend north and south from the equator to about 15° to 25° latitude. In these climates all months have average temperatures greater than 64°F (18°C) and annual precipitation greater than 59".

B - Dry Climates
The most obvious climatic feature of this climate is that potential evaporation and transpiration exceed precipitation. These climates extend from 20°-35° North and South of the equator and in large continental regions of the mid-latitudes often surrounded by mountains.

C - Moist Subtropical Mid-Latitude Climates
This climate generally has warm and humid summers with mild winters. Its extent is from 30° to 50° of latitude mainly on the eastern and western borders of most continents. During

the winter, the main weather feature is the mid-latitude cyclone. Convective thunderstorms dominate summer months.

D - Moist Continental Mid-Latitude Climates
Moist continental mid-latitude climates have warm to cool summers and cold winters. The location of these climates is poleward of the "C" climates. The average temperature of the warmest month is greater than 50°F (10°C), while the coldest month is less than -22°F (-30°C). Winters are severe with snowstorms, strong winds, and bitter cold from Continental Polar or Arctic air masses.

E - Polar Climates
Polar climates have year-round cold temperatures with the warmest month less than 50°F (10°C). Polar climates are found on the northern coastal areas of North America, Europe, Asia, and on the land masses of Greenland and Antarctica.

H - Highlands
Unique climates based on their elevation. Highland climates occur in mountainous terrain where rapid elevation changes cause rapid climatic changes over short distances.

Figure 5.1 is a map showing where these major categories occur in the mainland of the United States. If you want to see a color version of this map, you can find it at this link: https://www.weather.gov/jetstream/climates

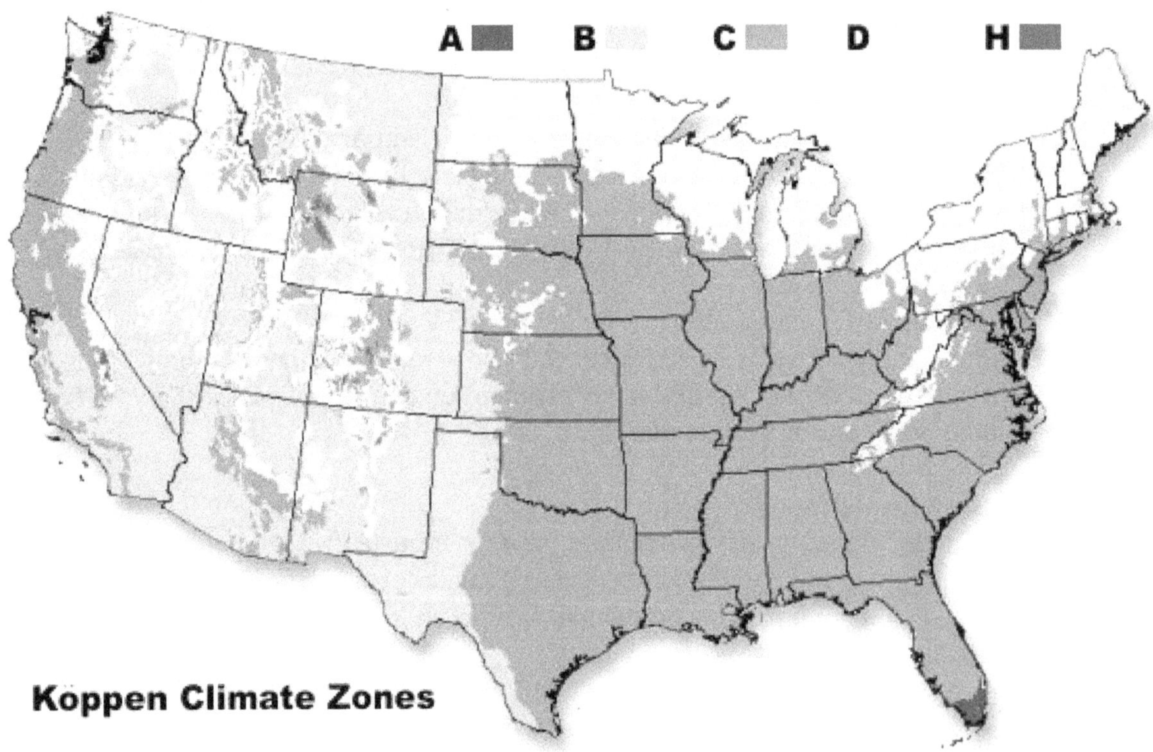

Figure 5.1 Koppen climate zones

The U.S. Department of Energy (DOE) has classified the climate of the United States into five zones: hot-dry/mixed-dry, hot-humid, mixed-humid, cold/very cold, and marine.

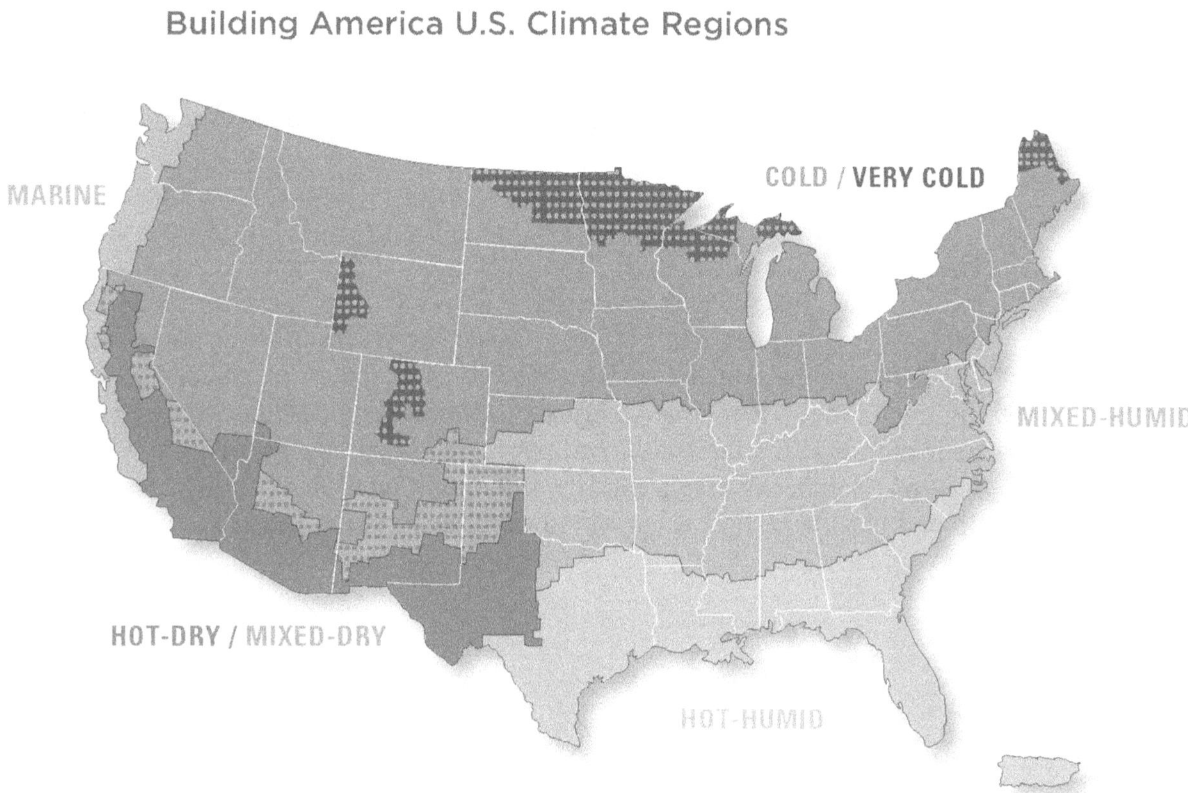

Figure 5.2 Department of Energy climate zones (Source: DOE)

You can also find county-by-county climate information at the following link:
https://www.energy.gov/eere/buildings/maps/building-america-climate-specific-guidance-image-map

1) **Hot-Dry/Mixed-Dry**

   The hot-dry and mixed-dry climates are defined as follows:

   - A hot-dry climate generally receives less than 20 inches (50 cm) of annual precipitation and the monthly average outdoor temperature remains above 45°F (7°C) throughout the year.
   - A mixed-dry climate receives less than 20 inches (50 cm) of annual precipitation, has approximately 5,400 heating degree days (50°F basis) or less, and the average monthly outdoor temperature is above 45°F (7°C) except during the winter months.

   DOE provides a free 187-page publication entitled "Builders Challenge Guide to 40% Whole-House Energy Savings in the Hot-Dry and Mixed-Dry Climates (Volume 9: Building America Best Practices Series)." Look through the PDF a few times and read the information that you are not familiar with. This should be adequate. The PDF can be found at the following link:
   https://www.energy.gov/sites/prod/files/2013/11/f5/18899.pdf

If it is possible to consider orientation at the site development level of the homebuilding process, one virtually no-cost option for improving energy performance is to subdivide for solar orientation.

Trees, houses, and community buildings can be positioned to plan for potential solar panel installations. Where community and private pools are installed in new construction, solar water heating systems should be included as part of the initial design.

Convection is the movement of heat via a gas or liquid. Warm air becomes buoyant, while cold air tends to sink. Convection is the force that draws hot air up a chimney. This force is sometimes called the stack effect. The ability of the air to carry water vapor is directly related to how warm it is. Warmer air can carry much more vapor than cooler air. Designers and builders can use convection as a natural way to cool and ventilate a home, but it can cause problems in the wrong places and circumstances. Figure 5.3 shows Building America's Building Science Consortium used convection to help cool this structural insulated panel (SIPs) cottage in Georgia. Cool air from the screened porches is drawn into the building's interior where it heats up and is pulled up and out the second-story windows, drawing more air in through the shaded first-story windows.

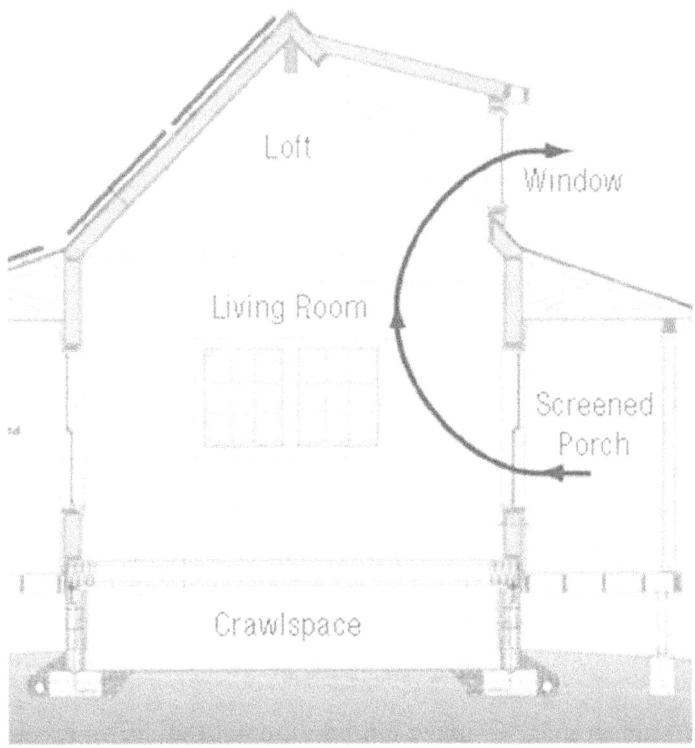

Figure 5.3 Convection (Source: DOE)

Other planning and design strategies include the following:

- Provide shading from direct sun.

- Use material with high thermal mass to store energy during daytime, and release the stored energy at night.
- Use pools to reduce local air temperature through evaporation.
- Use roof ponds for one- or two-story buildings to provide evaporation and thermal mass.
- Use compact forms with small surface areas.
- Use light exterior colors.
- Minimize the size of openings.

2) Hot-Humid

The hot-humid climate is defined as a region that receives more than 20 inches (50 cm) of annual precipitation and where one or both of the following occur:

- A 67°F (19.5°C) or higher wet bulb temperature for 3,000 or more hours during the warmest six consecutive months of the year; or
- A 73°F (23°C) or higher wet bulb temperature for 1,500 or more hours during the warmest six consecutive months of the year.

The Building America hot-humid climate region corresponds to zones designated 1, 2, and 3 (below the warm-humid line) on the International Energy Conservation Code (IECC) map (DOE 2010). You can find the color map at the following link: https://basc.pnnl.gov/images/iecc-climate-zone-map

There are nine states that are wholly or partially in the hot-humid Building America climate region of the United States, which stretches from Texas to Florida along the Gulf Coast, and from southern Florida to North Carolina along the Atlantic coastline.

Builders in the hot-humid climate must be able to address large solar gains in the summer and significant levels of moisture in the ambient air most of the year, along with torrential downpours and high winds, including hurricanes.

Passive and active solar design elements can be incorporated into homes throughout the hot-humid climate. Design considerations include lot orientation, roof tilt, overhangs, windows, and shading.

Even if the builder does not plan to put solar panels on the roof at initial construction, homes should be designed and sited with solar panels in mind so that homeowners can take advantage of this energy resource in the future. Significant amounts of unobstructed roof area are needed. Roof space that faces due south is ideal but not necessary. In the hot-humid climate, the optimal tilt angle or roof pitch for solar panels varies from 24 degrees along the Gulf Coast to 30 degrees in South and North Carolina. However, much of the solar energy available can still be collected even when the panels face nearly due east to nearly due west (DOE 2007)

The heat and humidity and the forces of driving wind and rain can take their toll on buildings in the hot-humid climate. The following are design considerations for the Hot-Humid Climate:

- Build slab-on-grade foundations and grade lots to drain away from the structure. In areas of likely coastal flooding, build on concrete or pressure-treated piers.
- Create a tight thermal envelope and install a positive pressure ventilation system.
- Place the air handler and ducts in conditioned space or go ductless with mini-split heat pumps.
- Install a reflective roof and use light or reflective exterior wall colors.
- Install a radiant barrier in the attic.
- Install high-performance, low-emissivity windows with low solar heat gain coefficient.
- Install overhangs, covered porches, awnings, pergolas, or shade trees to minimize solar heat gain.
- Install a dehumidifier to control shoulder-season humidity.

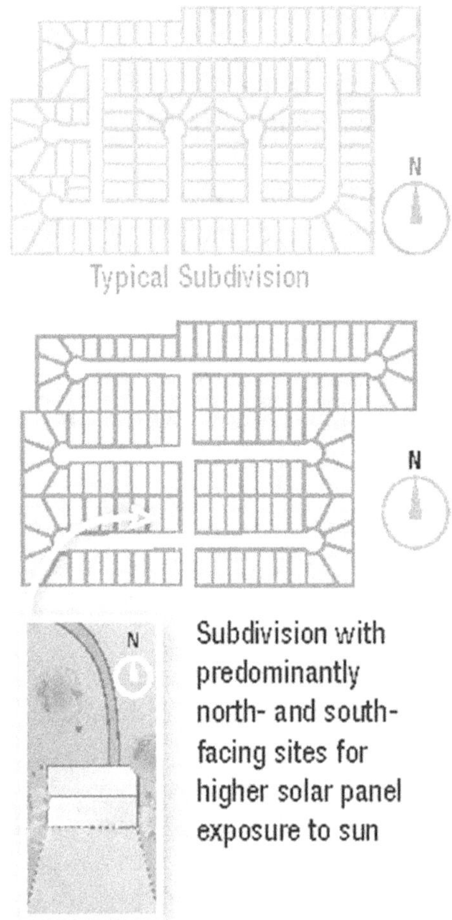

Figure 5.4 Lot Orientation: Plan subdivision lot lines and roads for predominantly north and south orientation (Source: DOE)

- Use paints that contain mildewcides.
- Use door jambs that are designed for water and rot resistance.
- Use cement backer board behind tubs and showers.
- Install a pest defense system in the walls.
- In framed homes, use borate pressure-treated lumber.
- Install a thermostat with humidity controls.

- Install outside showers at beach houses.
- Locate windows to catch coastal breezes.
- When installing windows use sill wrap, corner shields, and adhesive flashing tape to protect against water intrusion.
- Install hurricane strapping, rods, and bolts per code.
- On roofs use fully adhered roofing membrane at eaves and gable ends, butyl-based adhesive back flashing strips on roof sheathing joints, and hurricane-rated shingles. Use non-heat producing CFL lights.
- Install ceiling fans.
- In flood- and hurricane-prone areas, install a generator-ready electrical service panel to run generator-powered shopvacs, fans, and heaters to dry out the house and reduce water damage during post-storm recovery when electric power outages are common.
- In hurricane areas, consider steel-reinforced concrete walls, outswing doors, hurricane shutters, eave designs with extended fascia providing drip edge, and recessed soffit vents.

DOE provides a free 260-page, "40% Whole-House Energy Savings in the Hot-Humid Climate (Volume 15: Building America Best Practices Series)." You just need to look through the PDF a few times, and read the information that you are not familiar with. That should be adequate. You can find the PDF at the following link:
https://www.energy.gov/sites/prod/files/2013/11/f5/40percent_hot_humid.pdf

3) **Mixed-Humid**

The mixed-humid climate is defined as a region that receives more than 20 inches (50 cm) of annual precipitation, has approximately 5,400 heating degree days (65°F basis) or fewer, and where the average monthly outdoor temperature drops below 45°F (7°C) during the winter months.

The Building America mixed-humid climate corresponds to climate zones designated 4 (non-marine) and 3 in category A above the warm-humid line on the International Energy Conservation Code (IECC) map (DOE 2010). You can find the color map at the following link:
https://basc.pnnl.gov/images/iecc-climate-zone-map

Portions of the mixed-humid climate zones are subject to frequent and intense **rainstorms**, severe **thunderstorms**, and **hail**. Some areas are at high risk for tornadoes and high winds. Large portions of the Midwest and the South have been subject to flooding.

The mixed-humid climate is subject to **hurricanes** along the Eastern seaboard. The region is at low risk of earthquakes except for localized areas around Arkansas and South Carolina (USGS 2010).

Precipitation in the mixed-humid climate varies widely, from 20 inches in the drier parts of Kansas, Oklahoma, and Texas, to more than 50 inches in Tennessee and North Carolina.

Flooding is a concern along the coast and inland waterways of the mixed-humid climate. Large portions of the Midwest and South have been subject to flooding, with most of the counties of these areas experiencing four or more presidential disaster declarations due to flooding since 1965.

Builders should check state building codes and contact the local community floodplain administrator for information on local floodplain management regulations. Location-specific Flood Insurance Rate Maps (FIRMs) can be obtained at www.fema.gov/hazard/map/firm.shtm.

If a community adopts and enforces adequate floodplain management regulations, FEMA will make flood insurance available within the community. Title 44 of the U.S. Code of Federal Regulations (**CFR**) contains the National Flood Insurance Program (**NFIP**) criteria for floodplain management, including design and construction standards for new and substantially improved buildings located in special flood hazard areas (**SFHA**s) identified on the NFIP's FIRMs. An SFHA is an area that has a 1% chance of being flooded in a given year. [CFR Title 44, Section 60.3(a)(3) requires that all new construction in a flood-prone area be constructed with materials resistant to flood damage.] The lowest floor of a residential building must be elevated to or above the base flood elevation (**BFE**). All construction below the BFE is susceptible to flooding and must consist of flood damage-resistant building materials, for example brick; cement board; concrete; non-paper faced gypsum; steel or solid-wood preservative-treated studs, trusses, and joists; and ceramic tile.

FEMA (2009, 2011) and Building America (Lstiburek 2006) recommended that home builders do the following:

- Design homes on elevated foundations.
- Use strong connections between the foundation and the elevated building to prevent buildings from floating or washing off their foundations.
- Design drainable, dryable interior wall assemblies using non-water sensitive building materials such as masonry, spray foam, rigid fiberglass insulation, metal or pressure-treated lumber framing, and non-paper-faced gypsum, as shown in the figures below.

Every state in the mixed-humid climate includes areas in wind zone 4 (the highest risk zone on FEMA's scale of four U.S. wind zones) or areas designated as special wind regions. The mixed-humid climate is vulnerable to catastrophic **high-wind**, heavy-rain events, such as **tornadoes**. Proper structural fastening and impact-resistant windows, doors, and skylights are critical to surviving high winds. Proper use of roofing materials can help roofs withstand high winds and protect against severe rains. The following sources provide guidance for construction in high-wind areas.

The Federal Emergency Management Agency (FEMA), the National Association of Home Builders Research Center (NAHBRC), and the U.S. Department of Housing and Urban Development (HUD) have produced a series of 37 fact sheets for building homes in high-wind, high-rain coastal environments
www.fema.gov/rebuild/mat/mat_fema499.shtm

The guides cover siting recommendations; moisture barrier systems; housewrap, masonry, roof sheathing, and other building materials; door and window installation; roof and deck-to-wall flashing; tile and asphalt roofing techniques for high wind areas; foundations; and construction techniques in flood-prone areas.

The Insurance Institute for Business and Home Safety provides structural details, guidance, and a list of building materials acceptable for hurricane, high-wind, floods and other natural disasters. The Institute also sponsors a Fortified for Safer Living program that certifies homes that meet a package of "code-plus" upgrades that greatly increase a new home's resistance to natural perils, including hurricanes, tornadoes, wildfires, floods, freezing weather, hail and earthquakes, as well as fires and interior water damage (IBHS 2008). Insurance credits for certified homes are available in some states. See the Institute's website for more details, www.disastersafety.org.

In climates with high humidity, unwanted moss and algae can grow on roofs—especially on north- and east-facing surfaces where they can stain and damage most roofing types. To inhibit moss and algae growth, roofs must have sunlight and airflow, allowing them to dry properly. Options for inhibiting moss growth in heavily treed areas include using metal roofs, installing metallic zinc or copper strips along the roof just beneath the peak, or using copper-treated asphalt shingles. The better option from an environmental (and cost) standpoint is to allow adequate air and light to reach the roof by planting and trimming trees so that branches do not overhang the roof.

DOE provides a free 258-page, "40% Whole-House Energy Savings in the Mixed-Humid Climate (Volume 16: Building America Best Practices Series)." You just need to look through the PDF a few times, and read the information that you are not familiar with. That should be adequate. You can find the PDF at the following link:
https://www.energy.gov/sites/prod/files/2013/11/f5/40percent_mixed_humid.pdf

4) Cold/Very Cold

There are 34 states that are wholly or partially in the cold or very cold Building America climates. Conditions vary widely across these climate zones, which stretch from Alaska to Arizona and California to Rhode Island.

Builders in the cold climate must be able to address solar gains in the summer and extreme, prolonged low temperatures and snow accumulation in the winter, along with torrential downpours, high winds, and, especially in areas east of the Mississippi, high humidity. The recommendations in this Best Practices guide apply to the cold and very cold climate regions. If you aren't sure that your project is within these climate regions, check the Building America Best Practices Volume 7.1 Guide to Determining Climate

Regions by County (DOE 2010, www.buildingamerica.gov) for a list of counties and climate zones. The Building America cold/ very cold climate zones correspond to the International Energy Conservation Code (IECC) climate zones 5, 6, and 7. You can find the color map at the following link:
https://basc.pnnl.gov/images/iecc-climate-zone-map

A **cold climate** is defined as a region with between 5,400 and 9,000 *heating* degree days (65°F basis). The Building America cold climate corresponds to the IECC climate zones 5 and 6.

A **very cold climate** is defined as a region with between 9,000 and 12,600 *heating* degree days (65°F basis). The Building America very cold climate corresponds to IECC climate zone 7.

A **subarctic climate** is defined as a region with 12,600 *heating* degree days (65° basis) or more. The only subarctic regions in the United States are found in Alaska, which is not shown on the climate maps. The Building America subarctic climate zone corresponds to IECC climate zone 8. While some of the guidelines in this document would apply to construction in the subarctic, the subarctic climate has its own unique considerations, which are not addressed here.
The extreme temperature variations, forces of driving wind and rain, snow accumulation and melting can take their toll on buildings and contribute to premature aging. Based on work with its cold climate case study builders, Building America makes the following recommendations for designing homes in the cold and very cold climate zones where high snow accumulation is likely:

- Don't put balconies, stairs, parking, or entrances under eaves.
- Design to avoid ice dams. Place any necessary roof penetrations high toward the ridge.
- Avoid electric snowmelt systems in roofs.
- Avoid use of gutters in high snow load locations.
- Orient buildings to allow for southern exposure to entrances, drives, and walks to encourage snow and ice melting and reduce the need for snow and ice removal and snow storage.
- Protect entrances and sidewalks by locating them under overhangs or at gable ends.
- Anticipate wind direction and drifting.
- Use low-maintenance, high-durability exterior products – for example cement fiber board siding, composite decking materials, and stained cedar railings.
- Keep furnaces and ducts out of attics and air seal ceiling lid well to lessen likelihood of ice dam formation.
- Heat and insulate basements and crawlspaces.
- Place footings below frost depth.

DOE provides a free 235-page, "40% Whole-House Energy Savings in the Cold and Very Cold Climates (Volume 12: Building America Best Practices Series)." You just need to look through the PDF a few times, and read the information that you are not familiar with. That should be adequate.

You can find the PDF at the following link:
https://www.energy.gov/sites/prod/files/2013/11/f5/cold_climate_guide_40percent.pdf

5) **Marine**

The marine climate covers a narrow band paralleling the West Coast from the Canadian border south to the county boundary separating Ventura and Los Angeles counties in California. In some stretches, this band is only one county deep inland from the Pacific Ocean. The marine climate was designated in recognition of the mild temperatures and moist conditions found along the coast. However, the marine climate borders on the cold climate in the north and the hot-dry climate in the south and the more extreme conditions of these neighbors are found in some inland areas. Homes in the marine climate are faced with high levels of moisture, often in the form of rain, fog, or snow.

The marine climate is defined as a region that meets all of the following criteria:

- a mean temperature of the coldest month between 27°F (-3°C) and 65°F (18°C)
- a warmest month of mean less than 72°F (22°C)
- at least 4 months with mean temperatures more than 50°F (10°C)
- a dry season in summer. The month with the heaviest precipitation in the cold season has at least three times as much precipitation as the month with the least precipitation in the rest of the year.

The marine climate is at low risk for hurricanes, tornadoes, and severe thunderstorms. However, the region is vulnerable to frequent precipitation, high winds, seismic events, and forest fires.

Rain falls frequently in the Pacific Northwest. Sometimes it comes as wind-driven sheets, often a fine drizzle, and sometimes a drenching downpour, but rain-imbued weather patterns in the Northwest can settle into the landscape and last weeks, and sometimes months, at a time.

Precipitation in the marine climate varies from an average of 10 to 35 inches annually along the California coast south of San Francisco, to 30 to 50 inches annually north of San Francisco and along the Oregon and Washington coast with localized areas experiencing up to 200 inches annually. While the marine climate's population centers receive an average amount of annual rainfall compared to other U.S. cities, they are among the rainiest in terms of days of rainfall per year, with Portland, Seattle, Tacoma, and Olympia averaging 150 to 160 rainy days annually.

Flooding is a concern along the coast and inland waterways of the marine climate region. Builders should check state building codes and contact the local community floodplain administrator for information on local floodplain management regulations. Location-specific Flood Insurance Rate Maps (FIRMs) can be obtained at www.fema.gov/hazard\map\firm.shtm

Stormwater management is a significant issue in many marine climate cities. The city of Portland, Oregon, recommends controlling stormwater onsite by planting new trees; retaining existing trees; constructing swales, drywells, soakage trenches, planter boxes, vegetated infiltration basins, and flow-through planters; and using pervious pavement and turf block.

DOE provides a free 230-page, "40% Whole-House Energy Savings in the Marine Climate (Volume 11: Building America Best Practices Series)." You just need to look through the PDF a few times, and read the information that you are not familiar with. That should be adequate. You can find the PDF at the following link:
https://www.energy.gov/sites/prod/files/2013/11/f5/marine_40_guide.pdf

See discussions on season, wind, sun, building orientation, and shading in Chapter Three.

2. **Topography**
Topography and grading is an important skill for an architect and will be tested in ARE exams. There are a few important concepts and skills in dealing with landforms and grading (topographic). These are spot elevations, contour lines, ridges, valleys, swales, uniform slopes, concave or convex slopes, grade of a slope, and rise and run of a slope, etc.

One tip to understand grading problems is that water always flows from a higher contour line to a lower contour line in a direction *perpendicular* to the *higher* contour line. (See arrow showing direction of water flow in figure 5.6 and figure 5.8.)

Contour lines are the most distinguishing characteristic of a topographic map. Contour lines are lines drawn on a map connecting points of the same elevation. If you physically followed a contour line, elevation would remain constant. Contour lines show the shape of the terrain and elevation.

Topographic maps show contour lines at even spacing. The spacing is called the **contour interval**. For example, if your map uses a one-foot contour interval, you will see contour lines for every one-foot change in elevation. These lines can occur at intervals of one, two, three, four, and so on. Depending on the topography, different maps use different intervals.

Every *fifth* contour line is an **index contour**, and the index contour lines are the only ones labeled. The index contours are drawn in a darker line.

Ridge and valley (or **swale**) are important concepts in grading. **Ridge** bottoms are expressed by "U" or "V" shaped contour lines with their closed end pointing towards the *lower* elevation. **Valley** bottoms are expressed by "U" or "V" shaped contour lines with their closed end pointing towards the *higher* elevation (figure 5.6).

Run is the horizontal distance between two contour lines. **Rise** is the vertical distance between two contour lines. Three ways to express a **slope** are ratio, percent, and angle. (See figure 5.5 and figure 5.7)

	Formulas	Example	Results
Slope ratio (H:V)	HD:VD	100:50	2:1
Slope percentage (%)	(VD ÷ HD)	(50 ÷100) x100	50%
Slope angle (°)	ArTan (VD ÷ HD)	ArTan (50 ÷100)	26.6°

HD: Horizontal distance
VD: Vertical distance

Figure 5.5 Slope determination formulas

The most commonly used roof pitches (roof slopes) are between 4/12 and 9/12.
2:1 is the maximum slope allowed in a site design without using a retaining wall.

You can use Google to find sample contour line examples and study how they look. The following link is a good example:
https://adventure.howstuffworks.com/outdoor-activities/hiking/how-to-read-a-topographic-map2.htm

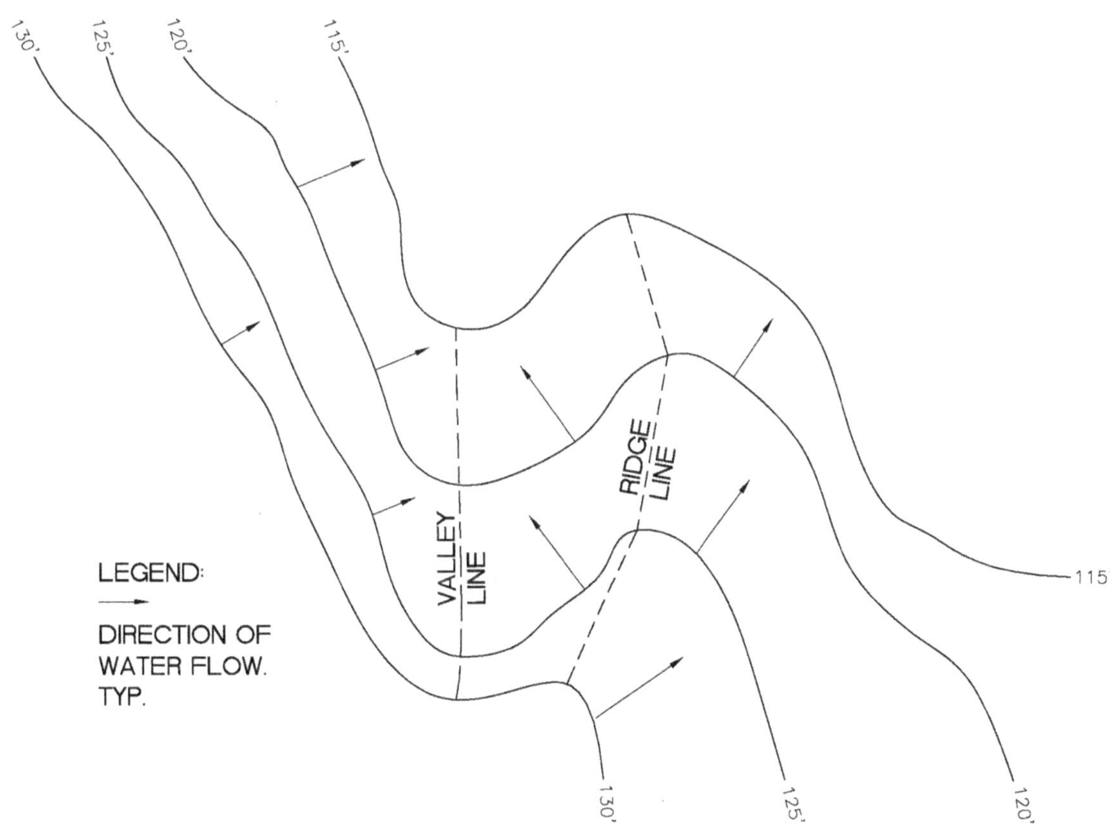

Figure 5.6 Ridge and valley

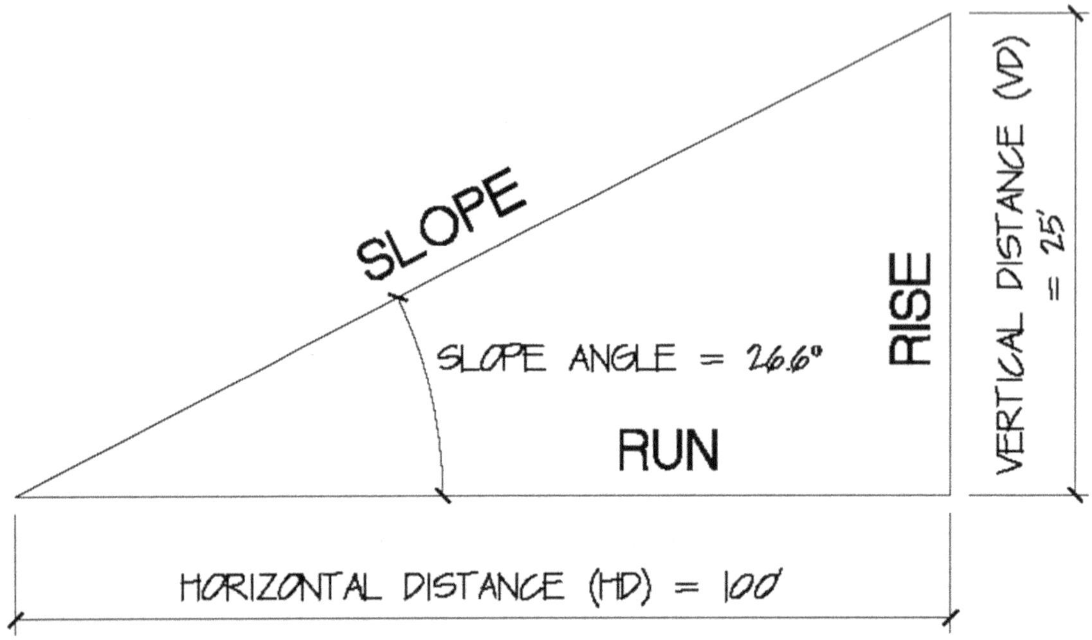

Figure 5.7 Rise, run, and slope

Spot elevations are used to show the elevations of some key points on a topographic map or grading plan, such as the elevations at the corners of a building. Check the building pad elevation to verify that the building is raised high enough to avoid flooding problems on the first floor. Spot elevations are also used at entrances or exits of a building to check the accessibility of a building's entrances or exits.

Check structural plans and make sure *all* deepened footing and retaining conditions at the perimeter of the building are shown. Coordinate with the architectural floor plan and civil grading plan for grades adjacent to the building.

Note:
You as an architect should send the civil grading plan to the structural engineer at the beginning of the project (or as soon as it is available) for his coordination.

It is important to coordinate with civil plans: boundary and topographic survey plans for property lines and easement, horizontal control plans for dimensions, grading plans for grades and handicap (accessible) path of travel, drainage plans for storm drains and catch basin locations and points of connections (P.O.C.) and concrete swales, curbs, and gutters, and utility plans for P.O.C.s for utilities (gas, water, fire sprinkler lines, sewer lines, etc.). Provide utility markers at all utility line stub-up locations.

On architectural site plan, use keynotes to indicate that the contractor should provide rough grading and compaction for the entire proposed building area per soils report's recommendation.

After reviewing figure 5.8, can you tell which one is a concave slope?

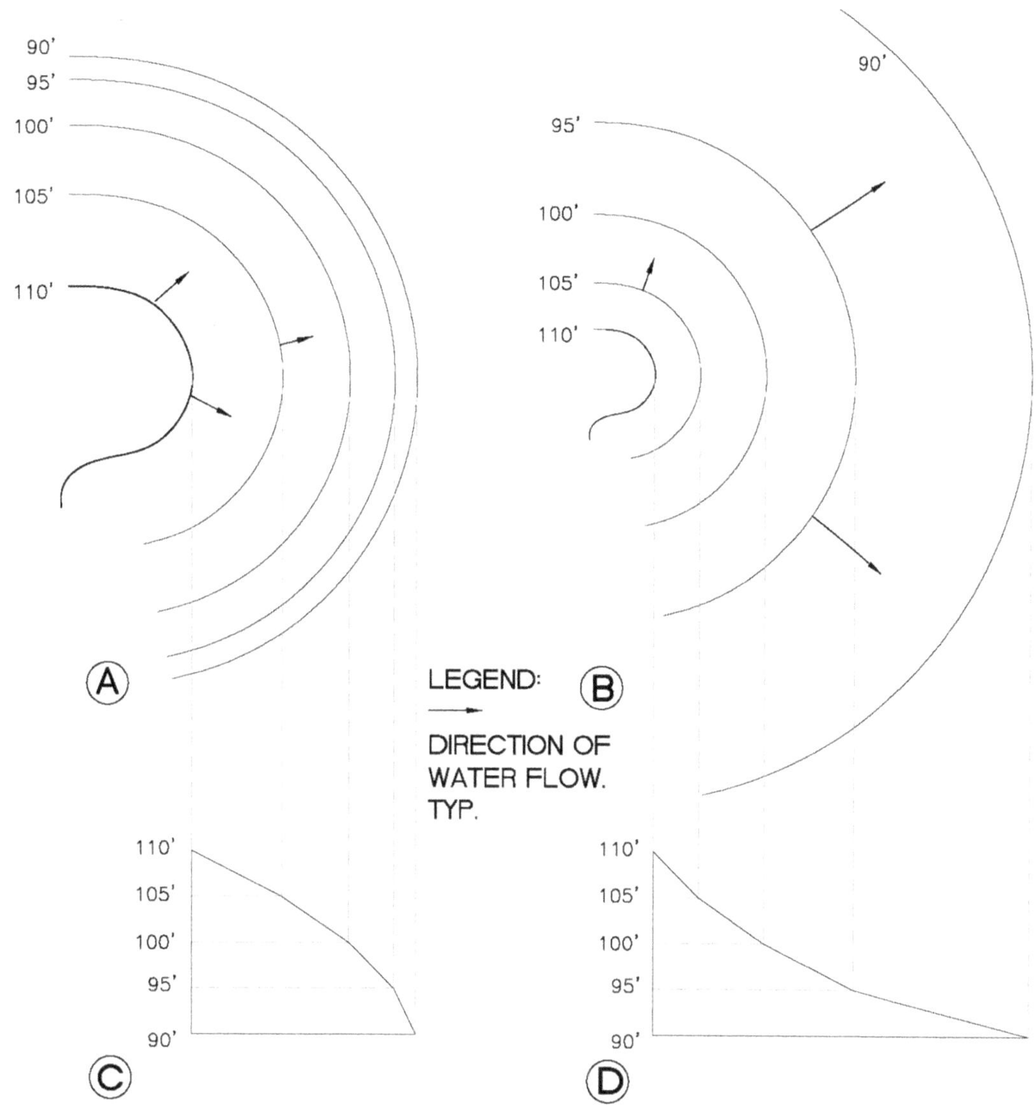

Figure 5.8 Concave or convex slopes

The following are some tips on how to differentiate a concave slope from a convex slope: For a ***concave*** **slope,** the contour lines are closer to each other at the *higher* portion of the slope. So, B and D in figure 5.8 represent a concave slope.

For a ***convex*** **slope,** the contour lines are closer to each other at the *lower* portion of the slope. So, A and C in figure 5.8 represent a *convex* slope.

It is very easy to confuse the term convex slope with the term concave slope. Here is a simple **mnemonic** to help you differentiate them.

A *concave* slope contains a cave and has a "dent."

Earthwork includes excavation and grading. **Excavation** is the removal of soil to allow construction of a foundation or other features below the finished grade. A long and narrow excavation for piping or electrical conduits or a narrow foundation wall or footing is called **trenching**.

Excavation can be dangerous to workers. Unshored soil has a slope steeper than its natural angle of repose, and it will collapse. Temporary **shoring** or **bracing** may be required for a steep slope during excavation and construction. They are typically the responsibility of the contractor, and not included in a typical structural engineer's scope of work.

Excavation is also called **cut**, and back fill is also called **fill**. It is a good practice to balance the cut and fill of a site to minimize the need to bring in extra soil from an off-site location, or to transport soil from the site to a landfill.

Grading is modifying the site contours per the grading plan. **Rough grading** is moving the soil to within about six inches to one foot of the desired grades. **Finish grading** is moving the soil to within one inch of the desired grades, and often includes the placement of the topsoil.

ARE exams have evolved to ARE 5.0 and will keep on evolving, but some of the fundamental knowledge has remained the same. The grading vignette for ARE 4.0 is a very good example of how to do grading using cut and fill and swales. **Cut** means removing soils from the site**, fill** means adding soils to the site. We will briefly discuss it here.

Figure 5.9 shows a proposed building pad with existing contour lines.

Figure 5.10 shows new contours with solid lines, and existing contours with dashed lines. The grading solution depicted basically cuts into existing contours to create two separate **swales** (or valleys) around a proposed building pad which drain water away from the building pad. The building pad elevation is set at 106'-6", and the elevation for the highest contour line near the building pad is six inches below the building pad at 106'. The contour lines are shown at sharp angles because of NCARB software limitations. In real grading plans, we should always try to make the contour lines smooth.

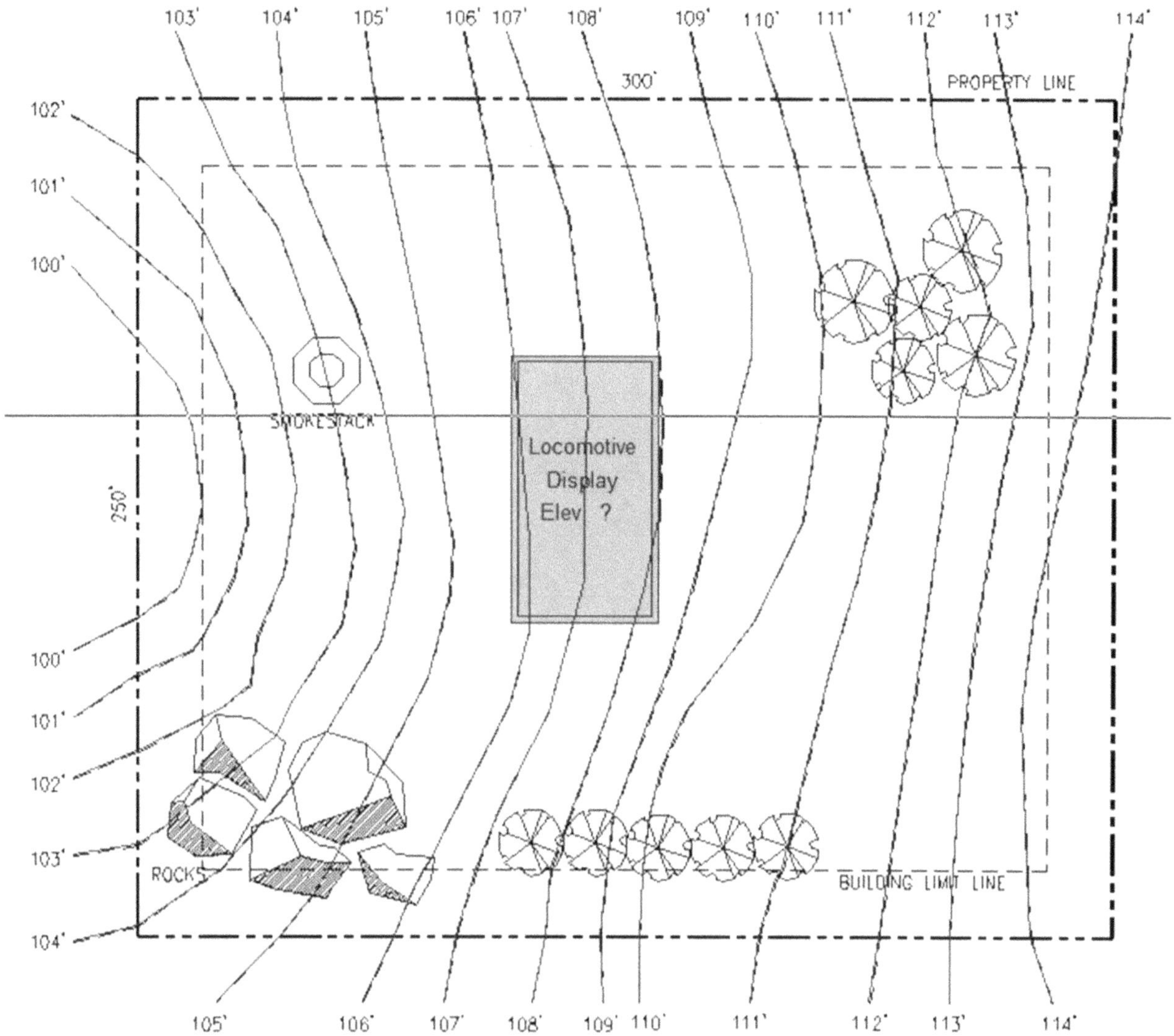

Figure 5.9 A proposed building pad and existing contour

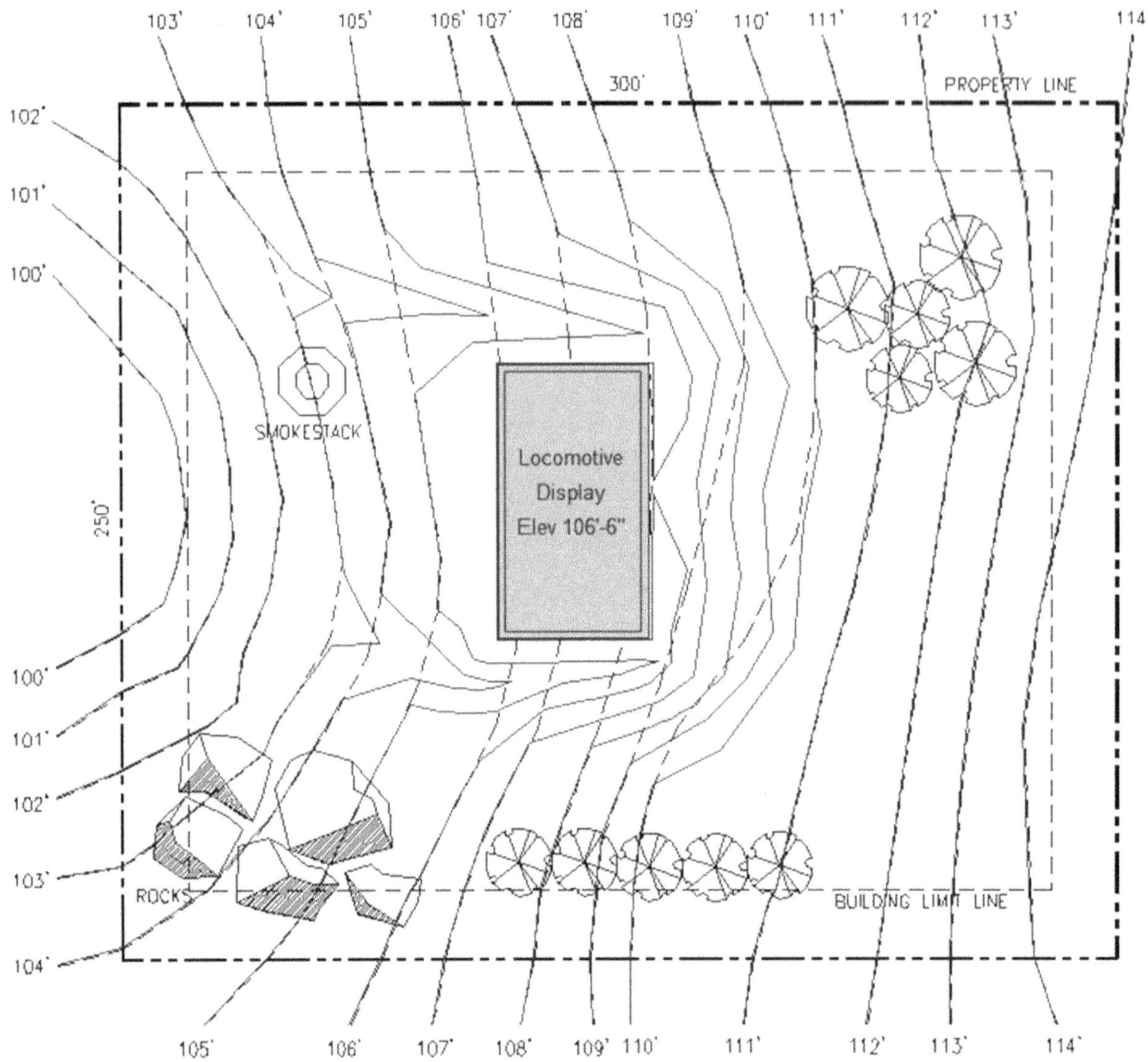

Figure 5.10 Creating swales around a building pad

3. Drainage

The two main categories of drainages are subsurface and surface.

Subsurface water can cause differential settlement, reduce soil load bearing capacity, and leak into buildings.

The **water table** is the level below which the ground is saturated with water. If a portion of a building is below water table, it is subject to additional loads, hydrostatic pressure, and poses extra difficulty in waterproofing.

To minimize subsurface water, the ground adjacent to a building must be drained away from the building, at a slope not less than ¼ inch per foot. Water from decks and roofs should be drained away from the building with gutters, pipes, and splash blocks. French drains, storm drain systems, and drywells can be used for drainage below the ground. Gravels, open-web matting (geotextile material), and waterproofing membrane can be used to relieve water pressure against the slope or underground walls.

Surface water should be drained away from the building by modifying the contours around the building. Gutters can be designed and built as part of the curb to collect water from paved areas. Catch basins, drain inlets, manholes, and storm drains can be used for surface water drainage too.

EPA provides the following guideline for site drainage:
Guidance 1: The site drainage design creates a controlled condition to help move water away from the building. To the extent possible, the design maintains the rate of water-soil infiltration (i.e., the downward entry of water into the surface of the soil) at the site before the site was disturbed. Runoff (i.e., water that does not infiltrate into the soil) must be managed by other drainage methods.

Guidance 2: Avoid unnecessary impervious surfaces. Avoiding unnecessary or large impermeable surfaces—or using alternative, relatively permeable paving materials—will allow more water to infiltrate, thus reducing the size and cost of systems managing runoff. Placing facilities on a site changes the site's drainage characteristics by increasing the impervious area, which, in turn, increases the volume of runoff that must be managed. Where large expanses of impervious surface are unavoidable, such as parking lots, breaking the expanse into smaller areas or using alternative permeable pavement techniques can help reduce runoff.

Alternative paving materials such as pervious pavement, modular porous paver systems or other surfaces can be used to reduce runoff.

- Porous pavement is a permeable surface often built with an underlying stone reservoir that temporarily stores surface runoff before it infiltrates into the subsoil. Porous pavements may be made using asphalt or concrete. Medium-traffic areas are the ideal application for porous pavement. Porous pavement may be inappropriate in areas such as truck loading docks and areas where there is a great deal of commercial traffic.

- Modular porous pavers are permeable surfaces that can replace asphalt and concrete; they can be used for driveways, parking lots and walkways. Alternative pavers can replace impervious surfaces, resulting in less storm water runoff.

- The two broad categories of alternative pavers are paving blocks and other surfaces, including gravel, cobbles, wood, mulch, brick and natural stone.

Guidance 3: Use grading to slow down runoff and achieve a more balanced infiltration rate. Topography helps determine the amount, direction and rate of runoff. To the extent possible, retain existing contours so that the existing drainage patterns can be maintained. Grading also can be used to correct drainage problems. Where steep slopes contribute to rapid runoff, re-grading to more moderate slopes can reduce runoff velocity.

Guidance 4: Ensure positive site drainage principles are met, including:

- Making certain water is moved away from the building.
- Ensuring water is not allowed to accidentally pond in low areas.
- Making sure the finished floor is elevated enough so that water will not back up into the building if the drainage systems are blocked.

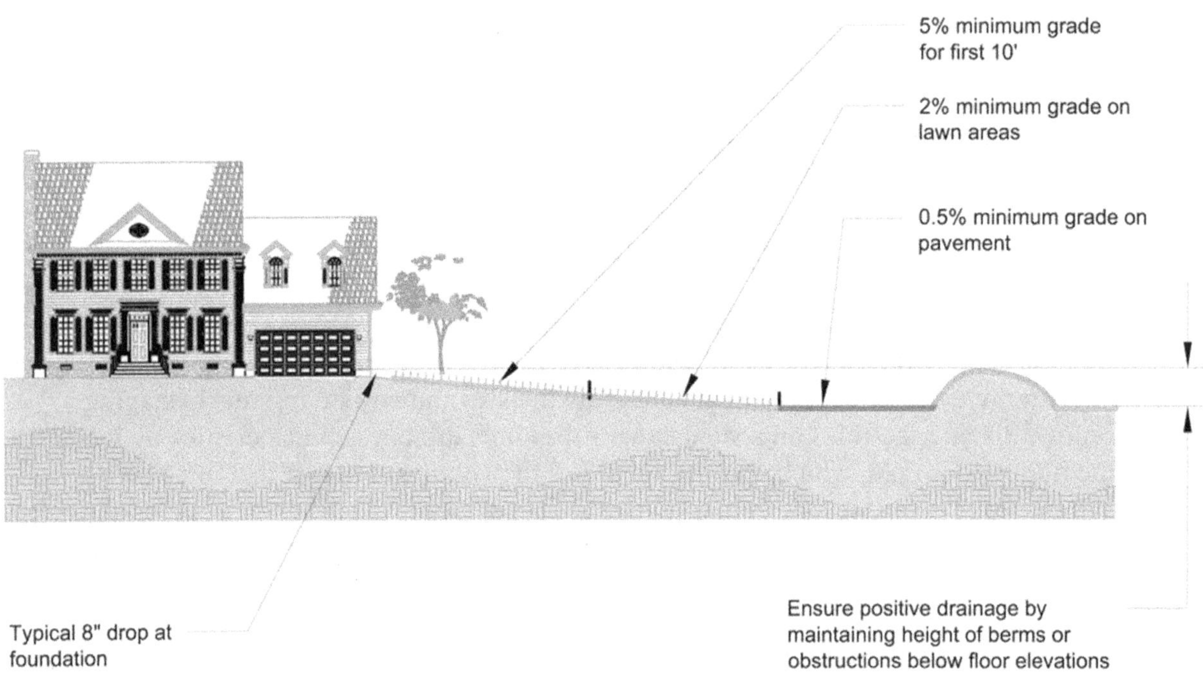

Figure 5.11 Positive drainage principles

Guidance 5: When runoff must be controlled and redirected away from the building, identify and design water runoff management approaches appropriate for the site's characteristics. Potential approaches to use include:

Infiltration control methods such as swales or infiltration trenches.

- A swale (i.e., a grassed channel, dry swale, wet swale, biofilter or bioswale) is a vegetated, open-channel management practice designed specifically to treat and attenuate runoff for specified water quality and volume. As water flows along these channels, vegetation slows it to allow sedimentation; the water filters through a subsoil matrix or infiltrates the underlying soils or both.

- An infiltration trench (i.e., infiltration valley) is a rock-filled trench with no outlet that receives runoff. The runoff passes through some combination of pretreatment measures, such as a swale and detention basin, and into the trench. Runoff is stored in the spaces between the stones in the trench and from there infiltrates through the trench bottom and into the soil. The primary pollutant removal mechanism of this practice is filtering through the soil.

Retention or detention control methods such as wet or dry ponds.

- Wet ponds—storm water ponds, wet retention ponds and wet extended-detention ponds—are constructed basins that contain a permanent pool of water throughout the year or at least throughout the wet season. Ponds treat incoming runoff by allowing particles to settle and algae to take up nutrients. The primary removal mechanism is settling, which occurs as runoff resides in the pond. Pollutant uptake, particularly of nutrients, occurs through biological activity. Wet ponds traditionally have been widely used as a storm water best management practice.

- Dry detention ponds—dry ponds, extended detention basins, detention ponds and extended detention ponds—hold runoff for some minimum time to allow particles and associated pollutants to settle. Unlike wet ponds, these facilities do not have a large permanent pool of water; however, they are often designed with small pools at the basin's inlet and outlet. Dry detention ponds also can contribute to flood control by providing additional flood water storage.

For detailed information including applicability, design criteria, limitations and maintenance requirements on these and many other site drainage methods, visit EPA's storm water management website.

Guidance 6: Landscape irrigation systems must be designed so that they do not spray the building or soak the soil next to the foundation. Consider hiring a qualified irrigation designer or irrigation consultant to design the system, keeping in mind these considerations:

- Spray heads and rotor heads spray water into the air. When designing spray systems consider wind conditions. Wind can carry airborne water beyond the area intended to be covered, and the sprinklers may spray the building or the soil around the foundation.

- Drip irrigation is a slow, even application of water through plastic tubing that delivers water directly to plants. Drip irrigation systems use less water than spray systems; however, they still can soak the ground around the foundation and cause moisture problems in a building.

- All irrigation systems, regardless of type, should be properly controlled and monitored. Timers should be installed to ensure the system shuts off. Water flow meters should be installed to measure the volume of water moving through the system. Regularly monitored meters can be a source of information about excessive water use due to timer problems or system leaks. Consider installing devices such as tensiometers or soil blocks to measure soil moisture.

Guidance 7: Ensure water draining from one building or site does not violate the good drainage of an adjacent building or site. This can happen when a building is constructed close to an existing building and dumps drainage water (e.g., roof, surface, etc.) onto or at the existing building, overwhelming its drainage features.

Guidance 8: Consider green building practices that minimize the need for irrigation or that capture rainwater for use in irrigation.

- Select trees, shrubs, ground cover and other landscaping elements based on their ability to grow well with little or no additional water. Such plants will minimize the use of water for irrigation.

- Explore the potential for capturing, diverting and storing rainwater for landscape irrigation, drinking and other uses. This approach can be used in all climates.

Guidance 9: Develop a construction-phase stormwater-management plan. The plan should address at a minimum:

- Methods for minimizing the potential for storm water runoff during construction.
- Methods to drain storm water from the site and away from the structure during construction.
- Methods for preventing building materials from getting wet.
- Methods for keeping the building or portions of the building dry during construction.
- Policies and methods for drying materials and the building if they become wet.
- Construction-phase storm water management supervisory roles and responsibilities.

For detailed information on construction-phase storm water management, visit EPA's storm water best management practices website.

Guidance 10: Develop guides covering the O&M of the storm water management system. The guides should include:

- The theory of operation of storm-water-management systems.
- Inspection procedures.
- Maintenance procedures and requirements.

For detailed information on post-construction storm water management, visit EPA's storm water best management practices website.

For the verification of site drainage:

- If storm water from the site is to be conveyed to a municipal separate storm sewer system (MS4), get a list of MS4 operator's requirements for the municipality. Give the list to the construction manager before construction begins.

- Provide the construction manager with a list of construction-phase critical details and an inspection schedule of the site drainage system, identifying the sequence of inspections, parties responsible for the inspections, and required documentation of the inspection results.

- Provide the construction manager and the building owner with a list of post-construction inspection and maintenance requirements for the site drainage systems.

EPA provides a free 144-page, "Moisture Control Guidance for Building Design, Construction and Maintenance." You just need to look through the PDF a few times, and read the information that you are not familiar with. That should be adequate. You can find the PDF at the following link:
https://www.epa.gov/sites/production/files/2014-08/documents/moisture-control.pdf

Criteria for Assigning Group Symbols and Group Names Using Laboratory Tests				**Soil Classification**	
				Group Symbol	**Group Name**
COARSE-GRAINED SOILS More than 50% retained on No.200 Sieve	Gravels More than 50% of coarse fraction on No. 4 Sieve	Clean Gravels Less than 5% fines	$Cu \geq 4$ and $1 \leq Cc \leq 3$	GW	Well-graded gravel
			$Cu < 4$ and/or $Cc < 1$ or $Cc > 3$	GP	Poorly graded gravel
		Gravels with Fines More than 12% fine	Fines classify as ML or MH	GM	Silty Gravel
			Fines classify as CL or CH	GC	Clayey gravel
	Sands 50% or more of coarse fraction passes No.4 sieve	Clean Sands Less than 5% fines	$Cu \geq 6$ and $1 \leq Cc \leq 3$	SW	Well-graded sand
			$Cu < 6$ and/or $Cc < 1$ or $Cc > 3$	SP	Poorly graded sand
		Sands with Fines More than 12% fines	Fines classify as ML or MH	SM	Silty sand
			Fines classify as CL or CH	SC	Clayey sand
FINE-GRAINED SOILS 50% or more passes the no. 200 Sieve	Silts and Clays Liquid limit less than 50	Inorganic	PI > 7 and plots on or above "A" line	CL	Lean clay
			PI < 4 and plots below "A" line	ML	Silt
		Organic	Liquid limit—oven dried < 0.75	OL	Organic clay
			Liquid limit—not dried	OL	Organic silt
	Silts and Clays Liquid limit 50 or more	Inorganic	PI plots on or above "A" line	CH	Fat clay
			PI plots below "A" line	MH	Elastic silt
		Organic	Liquid limit—oven dried < 0.75	OH	Organic clay
			Liquid limit - not dried	OH	Organic silt
HIGHLY ORGANIC SOILS				PT	Peat

Figure 5.12 Unified Soil Classification System (USCS)

First and/or second letters		Second letter	
Letter	**Definition**	**Letter**	**Definition**
G	gravel	P	poorly graded (uniform particle sizes)
S	sand	W	well-graded (diversified particle sizes)
M	silt	H	high plasticity
C	clay	L	low plasticity
O	organic		

Figure 5.13 Legend for Unified Soil Classification System (USCS)

4. **Soil**

Soils are classified per their grain sizes and are either organic or inorganic.
Gravel has particles over 2 mm in diameter.
Sand has particles from 0.05 mm to 2 mm in diameter. The finest particle is just visible to the eye.
Silt has particles from 0.002 mm to 0.05 mm in diameter. The grains can be felt as smooth but are invisible.
Clay has particles under 0.002 mm in diameter. The substance is sticky and plastic when wet, and floury and smooth when dry.

Soils are classified per the Unified Soil Classification System (USCS), and are divided into major divisions and subdivisions according to grain size and lab tests of physical characteristics. USCS provides standardized names and symbols for soils. See figure 5.12. The Unified Soil Classification System (USCS) is a soil classification system used in geology and engineering to describe the grain size and texture of a soil. The classification system can be applied to most unconsolidated materials, and is represented by a two-letter symbol. Each letter is described in figure 5.13 (with the exception of Pt).

Note:
You do not need to memorize figure 5.12 and figure 5.13. Just read them a few times in order to understand and become familiar. This level of study should be adequate for the exam.

If the soil has 5–12% by weight of fines passing a #200 sieve (5% < P#200 < 12%), both grain size distribution and plasticity have a significant effect on the engineering properties of the soil, and dual notation may be used for the group symbol. For example, GW-GM corresponds to "well-graded gravel with silt."

If the soil has more than 15% by weight retained on a #4 sieve (R#4 > 15%), there is a significant amount of gravel, and the suffix "with gravel" may be added to the group name, but the group symbol does not change. For example, SP-SM could refer to "poorly graded SAND with silt" or "poorly graded SAND with silt and gravel."

Water in soils can reduce the soil load bearing capacity, cause differential settlement, and create serious problems for foundations. Foundations below the groundwater line are subject to hydrostatic pressure, have additional loads, and are difficult to waterproof.

To decrease settlement, increase bearing capacity, or do both, soils can be treated in the following ways:

- **Compaction** of soil can provide the required base for the foundation. A **sheepsfoot roller** is used to compact large areas.
- **Densification** is a type of on-site compaction and can be achieved by dropping heavy weights, vibration, or pounding piles into the ground and filling the voids with sand.
- **Drainage** in a proper fashion can reduce hydrostatic pressure and increase the strength of the soil.
- **Fill** that is new and engineered soil can replace existing soil not suitable for building. The fill can be gravel, sand, soil, or other proper materials. There is an optimum relationship between the fill density and optimum moisture content. All fill has to be compacted, typically within 90% to 100% of the optimum **proctor density**. Moisture content should be within 2% to 4% of the optimum moisture content.

  Note:
  *The **Proctor compaction test** is named after Ralph Roscoe Proctor, who showed that the dry density of a soil for a given compaction depends on the amount of water that the soil contains during soil compaction. The test is a lab method of experimentally determining the optimal moisture content at which a given soil type will become most dense and achieve its maximum dry density.*

- **Geotextiles** are flexible, permeable fabrics and mattings. Geotextiles are typically synthetic, and used in drainage planes against foundation, landscape roofs, and silt fences, etc. They are also used to stabilize marginal soils under paving and engineered fill below footings, and as erosion control on steep slopes.
- **Mixing** a layer of gravel or sand in with existing soil can improve its bearing capacity.
- **Surcharging** is the preloading of the ground with fill materials to cause settlement and consolidation. After the required settlement is achieved, the fill material is removed. This method is often not feasible because of the time and cost needed.

Footings and foundations must be placed *below* the frost line to avoid the structure from lifting up because of the heaving and expansion of soils when they freeze.

Clays expand when they get wet and shrink when they dry. Foundations must be isolated away from expansive soils such as bentonite or other clays. Using caisson foundation piers, or piles, to bear on materials below the expansive soil may be a solution. Grade beams can span between the piers to support the building walls.

See geotechnical reports in Section B of this chapter for related information.

5. **Built and natural features**

An architect needs to know the site's built and natural features, in order to fully utilize the desirable features and minimize or block the undesirable ones. This information can be gathered by visiting the site, researching and reviewing the existing documents, and reviewing the survey map.

An architect should always take advantage of existing built and natural features. For example, an old existing tree can remain and benefit the occupant. A view to a desirable built or natural feature may need to be preserved or taken advantage of. A site analysis or perspective view analysis may be required.

Sites with high water tables may cause problem in excavation. Sites with a lot of rocks near the surface may be expensive, or difficult, or even impossible to develop.

Clients will typically provide the architect with site surveys, title reports, soil reports, etc. at the beginning of the project. If these documents are not available, the owner should hire a geotechnical engineer to produce a soil report, and a surveyor to survey the project site.

Note:

*A unit used frequently to note site areas in a site survey or site plan is **acre**, while your building areas are measured in s.f., so you have to convert site area from acres into s.f. constantly to calculate site building coverage, floor area ratio, parking count, landscape area ratio, etc. You need to memorize this important fact: **1 acre = 43,560 s.f. (Mnemonics:** 4 + 3 = 7, 5 + 6 = 11. So, if you can remember **7/11**, then you can remember **43,560 s.f.**)*

ALTA means **American Land Title Association.** It is the national trade association of the **abstract and title insurance industry** and was founded in 1907. An **abstracter** is the person who prepares a summary (or **abstract**) of public records relating to a title to a specific parcel of land. An **abstract** can also be defined as the history of the title.

ALTA's official website is:
http://www.alta.org/

ACSM is the **American Congress on Surveying & Mapping.** It is a nonprofit educational organization to advance the sciences of surveying and mapping and related fields and was founded in 1941.

ACSM's official website is:
http://www.acsm.net/

An ALTA survey is a boundary survey based on a set of minimum standards that are adopted by the **ALTA/ACSM.** It shows rights-of-way, easement, improvements, encroachments, encumbrances, and other elements related to land ownership. It is often used for commercial properties and meets a high standard. It is often done by a civil engineer or a surveyor hired by the owner directly.

ALTA surveys cannot be completed without a current title commitment. The surveyor needs title commitment for the legal description of encumbrances and the property.

CLTA is the **California Land Title Association**, a nonprofit corporation founded in 1907.

CLTA's official website is:
http://www.clta.org/

An ALTA title insurance is typically more expensive than a CLTA title insurance, because it covers all the standard items that CLTA covers (title to the real property, defects, liens, and encumbrances, marketable interests in the real property, lack of a right of access to and from the real property, and forgeries or failed conveyances in the chain of title), plus the following:

Matters that can be discovered by a survey, and easements, discrepancies, encroachments, or conflicts in boundary lines not shown by the public records.

A **title report** is a written description of the title status of a real property. It usually includes how the title is held and names of titleholders, property description, encumbrances, tax rates, and real property taxes due. **A preliminary report** is a title report provided when the report is ordered, and **a final title report** is an up-to-date title report provided at the time of recording. A title report can be prepared by an abstractor, an attorney, a title company, or an escrow company.

The Public Land Survey System (**PLSS**) is a way of subdividing and describing land in the United States. PLSS surveys, which are available for portions of land in 30 southern and western states, are made by the Bureau of Land Management (BLM). The PLSS typically divides land into 6-mile-square **townships**. Townships are subdivided into 36 one-mile-square **sections**. Sections can be further subdivided into **quarter sections**, **quarter-quarter sections**, or irregular government lots. PLSS was created to divide parcels of public land; it is not useful for the accurate location of points and should not be confused with coordinate systems like latitude/longitude, Universal Transverse Mercator (UTM), or the State Plane Coordinate System.

Note:
The ***UTM (Universal Transverse Mercator)*** *coordinate system divides the world into sixty north-south zones, each 6 degrees of longitude wide. UTM zones are numbered consecutively beginning with Zone 1, which includes the westernmost point of Alaska, and progress eastward to Zone 19, which includes Maine. If UTM ticks are shown on a USGS topographic map, the zone is indicated in the credit legend in the lower left corner of the map collar. (Source: USGS)*

For states that have Public Land Surveys:

- Almost all historical topographic maps (1884-2006) include PLSS tic marks or gridlines.

- US Topo maps published 2009-2012 do not include any PLSS data.
- US Topo maps published 2013-present have a Public Land Survey System layer that can be turned on and off.
- The National Map Viewer displays "BLM Public Land Survey System (PLSS)". Use the "Add Data" tool at the top of the map.
- For GIS users, PLSS data layers are available through the BLM's REST service endpoint. An online table shows PLSS Cadastral National Spatial Data Infrastructure (**CadNSDI**) data set availability.

Learn more at:
https://www.usgs.gov/

6. **Utilities**
Utilities include storm drains, sanitary sewers, water lines, natural gas, electricity, internet, telephone, cable TV, steam, etc. Before starting a design, an architect should find out the locations of and **points of connection (POC)** for all existing utilities, and should contact the appropriate utility company to confirm their requirements. An architect should also try to minimize the lengths of the utility lines.

Storm drains and sanitary sewers depend on gravity, and they should have priorities in the design. The invert of existing storm drains and sanitary sewers should be considered in the design to make sure the pipes have adequate slopes for drainage.

Per IPC Table 704.1, Slope of horizontal drainage pipe:

Size: Diameter of the pipe (inches)	Minimum slope (inch per foot)
2 ½ or less	1/4
3 to 6	1/8
8 or larger	1/16

Figure 5.14 Slope of horizontal drainage pipe

For example, pipes with a 2 ½ inch diameter or less, should have a minimum slope of at least ¼ inch per foot.

7. **Access points**
A site typically has pedestrian access, customer vehicle access, and service vehicle access.

Site design must consider the availability and location of public transit. Building entrance and main site amenities should be near public transit if possible.

The governing agencies may have specific requirements limiting the number of curb cuts for vehicle access, and also a minimum distance that the curb cut can be to any intersection.

An architect need to contact the governing agencies at the beginning of a project to confirm these requirements.

8. **Traffic patterns**
A **circulation plan** is a schematic projection or model of how people and/or vehicles flow through a given area based on past experience.

Pedestrian and customer vehicle access should be separated from service vehicle access. Service access may include moving vans, daily delivery service, and truck loading. Verify with the client and the local governing agencies for the loading dock and truck turning radius requirements.

Architects and planners can use a circulation plan to discover how to make future improvements to the system.

B. Synthesize site reports with other documentation and analysis (A/E)

NCARB Requirements
In addition to analyzing the attributes of the site, you will also need to review and interpret site documentation such as geotechnical reports, landscape reports, archaeological studies, utility surveys, topographic maps, demographics, traffic studies, environmental data, historic reports, and other site related reports. This is used to determine the feasibility of a project and verify the selection of site related consultants needed to execute the project.

1. **Geotechnical reports**
Soils reports, or "geotechnical reports," or "geotechnical investigation reports," are prepared by a licensed civil engineer or a geotechnical engineer experienced in soils engineering. They provide earth conditions affecting a building. Most new building constructions will require a soils report, especially in areas with low strength or expansive soils, projects on steep slopes, or FEMA floodplains or areas with high groundwater, projects where the foundation will be supported by fill or projects with a lot of grading.

Architects and structural engineers will need to review the soils report and *pay special attention to*, and follow, the **recommendations** on what kind of slab and footing to use, and what kind of base and moisture barrier to use under them. For example, a soils report may recommend a 4" slab with #3 rebar at 24" each way, over 2" sand over **15 mil** (thick) moisture barrier over another 2" of sand, and 12" wide minimum x 24" wide minimum continuous footing.

Note:
1. A "mil" equals to 1/1000 of an inch, 15 mil = 15 x 1/1000 = 0.015 inches.
2. A #3 rebar means a rebar with a 3 x ⅛" diameter, and a #5 rebar means a rebar with a 5 x ⅛" diameter, and so on.

In most cases, a soils report is completed before the developer can finalize the lease with the tenants.

Some clients, especially some national retail tenants, do *not* like to have sand over the moisture barrier, because they do not want to have moisture trapped in the sand between the moisture barrier and the concrete slab. They prefer recommendation like 5" slab with #3 rebar at 24" each way, over 15 mil moisture barrier over 6" of crushed rocks, and 12" wide minimum x 24" wide minimum continuous footing. Please note that if you are using this option, you need to use at least 15 mil thick instead of a 6 mil or 10 mil thick moisture barrier, otherwise it may be too thin and may be penetrated by the crushed rocks below.

The architect may need to go back to the geotechnical engineer or civil engineer to find out if the soils can be adjusted to meet the Client's project criteria.

The architect should try to get the client's criteria before the geotechnical investigation is done and forward the tenant's criteria to the geotechnical or civil engineer right away to avoid delay and confusion in the project.

Common methods used to obtain geotechnical data and create a soils report include **test pits, core samples, seismograph,** and **steel rods** driven into the soil to detect the presence of rock and test density.

Geotechnical tests typically fall into four categories:

- **Test pits** are dug either manually or with an excavator to the depth desired in order to reveal the sub-surface conditions.

- **Trenching** is similar to test pits except that in this case, the pit is elongated as a trench to establish how the sub-surface conditions change over different parts of the work site. A range of soil samplers can be used to extract test samples including hand-driven augers, shovels, split-spoon samplers, modified California samplers, and Shelby tube samplers.

- **Borings**, usually those with small diameters, provide the opportunity to physically remove soil or rock samples for testing. Borings provide the advantage of letting you 'see' the actual materials, but for certain types of soils, borings can disturb the soil conditions. Therefore, the samples extracted may not represent what the conditions will be like for supporting and building structures because it is unscientific and void of actionable data.

Generally, soil samples from the above tests are taken to a lab for evaluation.

- **In-situ testing** methods (in the situation, or at site) include penetration tests such as **standard penetration tests (SPT)**, which penetrate via drilling, and various **cone penetration tests**, which penetrate via direct push. These tests measure the physical properties of the subsurface soil directly, without removal. The tests also provide the advantages of generating a more accurate reflection of conditions underground as well as avoiding the needs of sending samples out for lab testing.

You can Google the key words "boring log samples" to become familiar with how the soils reports look and how to interpret them. The article at the link below is helpful. https://www.ntsafety.com/2016/09/06/engineers-corner-reading-soils-reports-and-bore-logs-from-a-contractors-perspective/

2. **Landscape reports**
A site survey map normally shows the locations and sizes of tree trunks. For a larger project, a landscape architect may be hired to do the site landscape planning and design. As part of their services they may generate a site landscape report.

3. **Existing conditions**
Existing conditions of a site can be obtained by:
    - site visits
    - research of existing documents
    - site survey plans including a boundary survey plan and a topography plan
    - using Google maps

4. **Utility surveys**
Utility surveys can be part of the survey map, and typical survey maps do show the point of connection of existing utilities. Some cities or utilities may have existing utility maps for the site or areas near the site already.

    Some utility companies also have a free service to come out and mark the locations of underground utilities before digging. For example, a homeowner can call 811 or visit call811.com to request the free service to mark underground utilities before digging. This information will help prevent accidental damage of underground utilities such as gas lines or electrical conduits as well as possibly save property or lives.

5. **Topographic maps**
Topographic maps include the existing topography from maps for the area around the site, and the actual topographic survey created by a surveyor for the project.

    USGC has free topographic maps for large portions of the United States. See the link: https://www.usgs.gov/products/maps/topo-maps

    See the discussion in section A of this chapter for more information.

6. **Demographics**
Demographics is the statistical study of populations, especially human beings. Demographic analysis can cover whole societies or groups with criteria based on education, nationality, religion, income level, and ethnicity.

Demographics can help an architect understand the client base and potential building users. The information can be a valuable tool to assist in the architectural design.

The following is a sample of demographics for the city of Irvine, California:
https://www.cityofirvine.org/about-irvine/demographics

7. **Traffic studies**
   A **traffic study** is a detailed analysis of a transportation system supported by data collection, which thoroughly defines the problem. The study consists of data collection, analysis of findings, and proposed solutions.

   Traffic studies are typically performed by traffic engineers or consultant and used by architects or planners as part of the site analysis and planning.

8. **Environmental data**
   Drivers, pressures, state, impact and response (**DPSIR**) model of intervention is a model for describing the interactions between society and the environment: For example, an interaction can be the human impact on the environment and vice versa.
   - **Driving forces:** e.g. tourism, industry, economic growth
   - **Pressures:** e.g. land-use change, pollution, population growth
   - **States:** e.g. soil quality, water quality, air quality, habitat, vegetation
   - **Impacts:** e.g. habitat fragmentation, ill public health, economic crisis, environmental damage, biodiversity loss
   - **Responses:** e.g. taxes, environmental laws

   Environmental data is based on "P", "S" and "I" of the DPSIR model: the measurement of environmental pressures, the state of the environment, and the impacts on ecosystems.

   All data generated by the execution of environmental law is considered environmental data. Environment statistics are usually generated by statistical offices. Environmental data has become more and more important to investors in recent years, prompting Bloomberg L.P. to begin providing environmental, social, and governance (ESG) data through their terminals.

   For an architectural project, environmental data is often covered in the **environmental impact statement (EIS)** or **environmental impact report (EIR)**. An EIS or EIR refers to the assessment of all possible impacts that a project may cause on the environment in an area. It includes traffic, population, schools, fire protection, archeological artifacts, endangered species, and community beauty.

   Governing agencies may require an EIR as a prerequisite for project approval.

   See the following link for more information:
   https://en.wikipedia.org/wiki/Environmental_impact_statement

9. **Historic reports**
   A site may have a historic structure or other features. Per National Park Services, a historic report provides documentary, graphic, and physical information about a property's history and existing condition. Broadly recognized as an effective part of preservation planning, a historic report also addresses management or owner goals for the use or re-use of the property. It provides a thoughtfully considered argument for selecting the most appropriate approach to treatment, prior to the commencement of work, and outlines a scope of recommended work. The report serves as an important guide for all changes made to a historic property during a project repair, rehabilitation, or restoration, and can also provide information for maintenance procedures. Finally, it records the findings of research and investigation, as well as the processes of physical work, for future researchers.

   See following link for more information:
   https://www.nps.gov/tps/how-to-preserve/briefs/43-historic-structure-reports.htm

10. **Other site related reports**
    Other site related reports include title reports, natural hazard disclosure reports, feasibility studies, financial reports, etc.

C. Analyze graphical representations regarding site analysis and site programming (A/E)

NCARB Requirements
You must be able to evaluate and understand diagrammatic graphics and how they are used to represent and communicate site conditions, relationships, and program requirements. These graphics include topographic, programming, circulation, adjacency, environmental, views, and vegetation diagrams, which communicate site information and relationships

1. **Topographic**
   See discussion in Section A of this chapter.

2. **Programming**
   Programming is the process of defining the problem and establishing the criteria or needs of the project. Function, form, cost, and time are the **four major concerns** of programming. Per *Problem Seeking: An Architectural Programming Primer* by William M Pena and Steven A. Parshall (Wiley, 2012), programming typically includes the following five steps:
   - **Clarifying goals:** What is the goal of the client? What do they want to achieve with these spaces? These answers suggest the physical means for the goal.
   - **Gathering information:** This information includes existing conditions and the client's requirements such as site conditions, number and characteristic of the users, space adjacency needs, equipment, growth rate, budget, code requirements, and climate.
   - **Forming concepts:** Abstract solutions that do not define physical and specific means are needed. These solutions later become the basis for design concepts.
   - **Finding needs:** Help the clients balance what they want and what they can afford, and find the real need of the client. The four elements of cost (quality, quantity, budget, and time) may need to be adjusted.

- **Stating the problem:** This statement bridges the gap between programming and the design process. The client and the architect agree on the real needs and the most important aspects of the problem, as well as set up criteria for design solutions.

Again, the ARE exam has evolved to ARE 5.0 so far and will keep evolving, but some of the fundamental knowledge has remained the same. We can use the ARE 4.0 site design graphic vignette as a sample of how to use graphics to do programming. (If you need to see the original requirements of ARE 4.0 site design vignette, you can easily find it on the NCARB website by simply searching on Google.)

- The Site Design vignette has a very complicated program. If you simply read the program several times, you may NOT cover all the program requirements when you layout the site design. It is also VERY hard to understand and summarize the program requirements by simply reading it multiple times.
- One effective way of digesting the program requirements is to do a simple **bubble diagram**: draw a simple **bubble diagram** to show the relationships between the site elements (figure 5.15). This is the *hardest* part.

Note:
*A **bubble diagram** is the most important tool for passing Site Design.*

- The bubble diagram does NOT have to be pretty and does NOT have to be to scale. The key is to do the bubble diagram quickly and turn the program requirements into a graphic in the shortest time. The bubble diagram is for *you*. As long as *you* can read it, it is good enough. Architects love graphic thinking. A bubble diagram can be a great tool.
- A **bubble diagram** can be a few simple hand-drawn circles: A circle with a label such as OT represents the Office Tower; R represents the Restaurant; PP represents the Pedestrian Plaza; WW represents a walkway; an arrow between two circle represents direct access required between the two elements; an arrow with a number above it represents a clearance distance between the two elements; a line connecting two elements indicates that they need to be placed close to each other; an eye symbol represents an element should be visible from a certain location (figure 5.15), etc.

Figure 5.16 shows the final solution for ARE 4.0 site design graphic vignette.

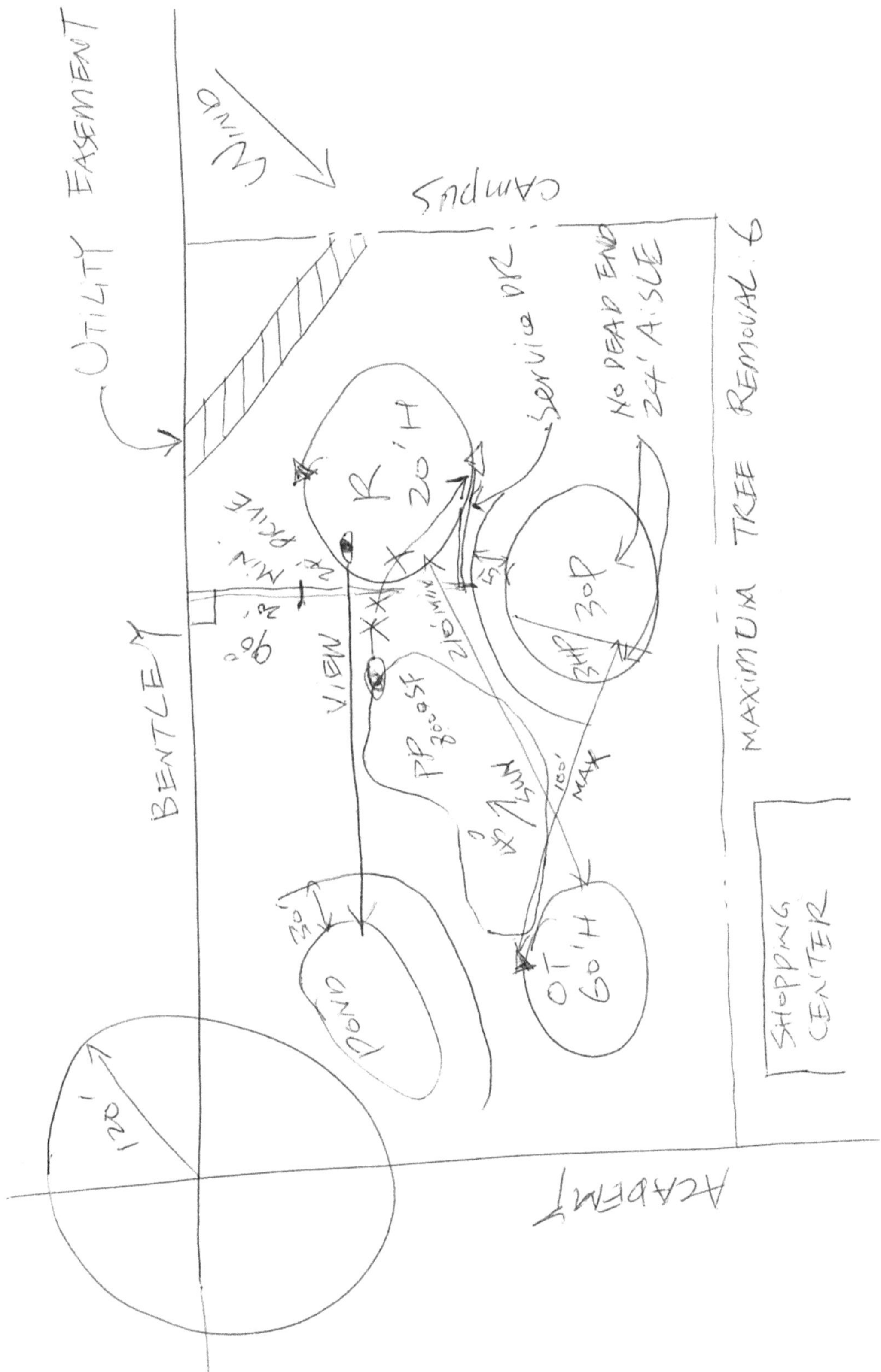

Figure 5.15 One effective way of digesting the program requirements is to do a simple **bubble diagram**

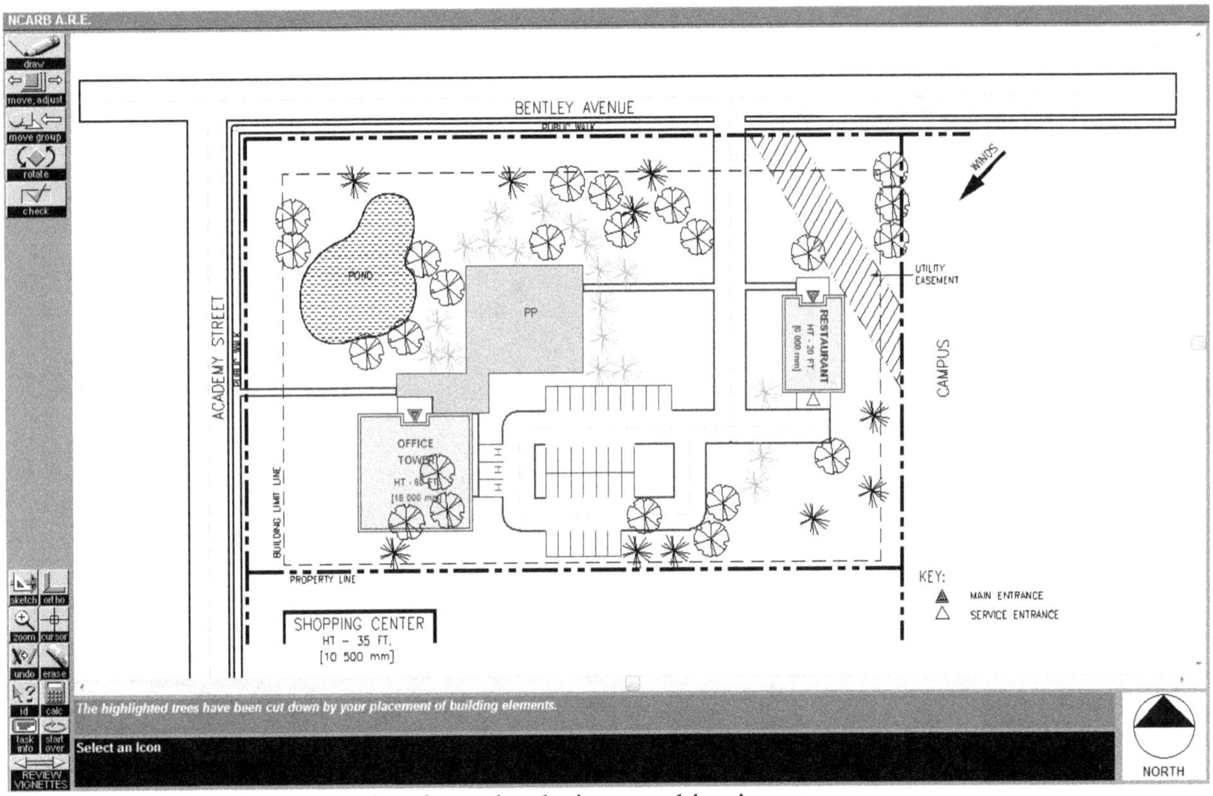

Figure 5.16 Final solution for a site design graphic vignette

3. **Circulation**

   Per the General Service Administration (GSA), site circulation design will vary greatly depending on the context, which can range from tight urban sites to suburban campuses or isolated rural settings. Yet the basic criteria remain the same in all situations. The site design should segregate pedestrian access, vehicular access (including parking), and service vehicle access at a minimum.

   - **Service Traffic:** Service dock access may be from an alley, from a below-grade ramp, or from a site circulation drive. If large trucks are to service the facility, sufficient maneuvering space must be provided, and the service drive shall be screened as much as possible. This type of traffic should always be separate from access to the parking garage. Where possible, a one-way design for service traffic is preferable to avoid the need for large truck turning areas. The service area of the facility shall not interfere with public access roadways.

   - **Public Transportation:** The project team should encourage the use of public transportation. The availability of a bus stop should be considered early in the design of a building in an urban setting and should be discussed with planners of the mass transit system. The project team should consider how to treat the orientation of the building and the site design and landscaping to encourage use of public transit and to address pedestrian traffic "desire lines" between the building entrance and transit stops.

- **Pedestrian Circulation:** The project team should consider neighboring uses, existing pedestrian patterns, local transit, and building orientation to anticipate pedestrian "desire lines" to and from the building from offsite. Designers should avoid dead ends, inconvenient routes, and the like. They should also consider how people moving across the site might help to activate sitting areas, outdoor art, programmed events, etc.

- **Drop-Off:** If the security analysis determines it is feasible, a vehicular drop-off area should be located on the street nearest the main entrance and, if site conditions permit, also near the entrance to the childcare center, if the project includes one.

- **Fire Apparatus Access:** Fire department vehicle access shall be provided and maintained to all new construction and alterations in accordance with the requirements of the National Model Fire Code that is used, NFPA 241, and NFPA 1141. The local fire department shall be consulted with regard to their specific requirements regarding the surface material of the access roadway(s), minimum width of fire lane(s), minimum turning radius for the largest fire department apparatus, weight of largest fire department apparatus, and minimum vertical clearance of largest fire department apparatus.

- **Vehicular Drives, Parking Lots and Service Area:** Follow local codes for entrance driveways within the right-of-way limits of city, county, or state-maintained roads.

- **Aerial Apparatus:** Buildings or portions of buildings exceeding thirty feet in height from the lowest point of a fire department vehicle access shall be provided with access roads capable of accommodating fire department aerial apparatus. Overhead utility and power lines shall not be within the aerial access roadway. In addition, at least one access road having a minimum unobstructed width of twenty-six feet shall be located within a minimum of fifteen feet and a maximum of thirty feet from the building. Also, at least one side of all buildings shall be accessible to the fire apparatus.

- **Surface Parking Lots:** Parking stalls are typically a minimum of 9'-0" wide and 18'-6" long, with two-way aisles of 24'-0". Where possible, ninety-degree parking should be used. Accessible parking spaces must be provided; these shall comply with the UFAS/ADA in quantity, location, and size.

  Internal islands for landscape planting should occupy no less than 10 percent of the total parking lot area. Curbs should be provided around the parking lot perimeter and around landscape islands.

  The maximum combined gradient for parking lots should not exceed 5 percent.

  For more information, see the GSA website at:
  https://www.gsa.gov/node/81661

4. Adjacency, adjacency matrix, and bubble diagram

Adjacency relationships are important for both site planning and building design. Some buildings or spaces must be adjacent to each other (**mandatory adjacency**), while others may need to be close to each other but not right next to each other (**secondary adjacency**), and some may need to be separated.

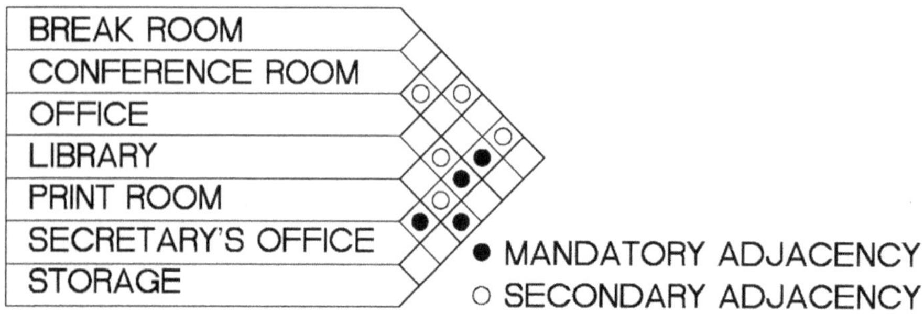

Figure 5.17 Adjacency matrix

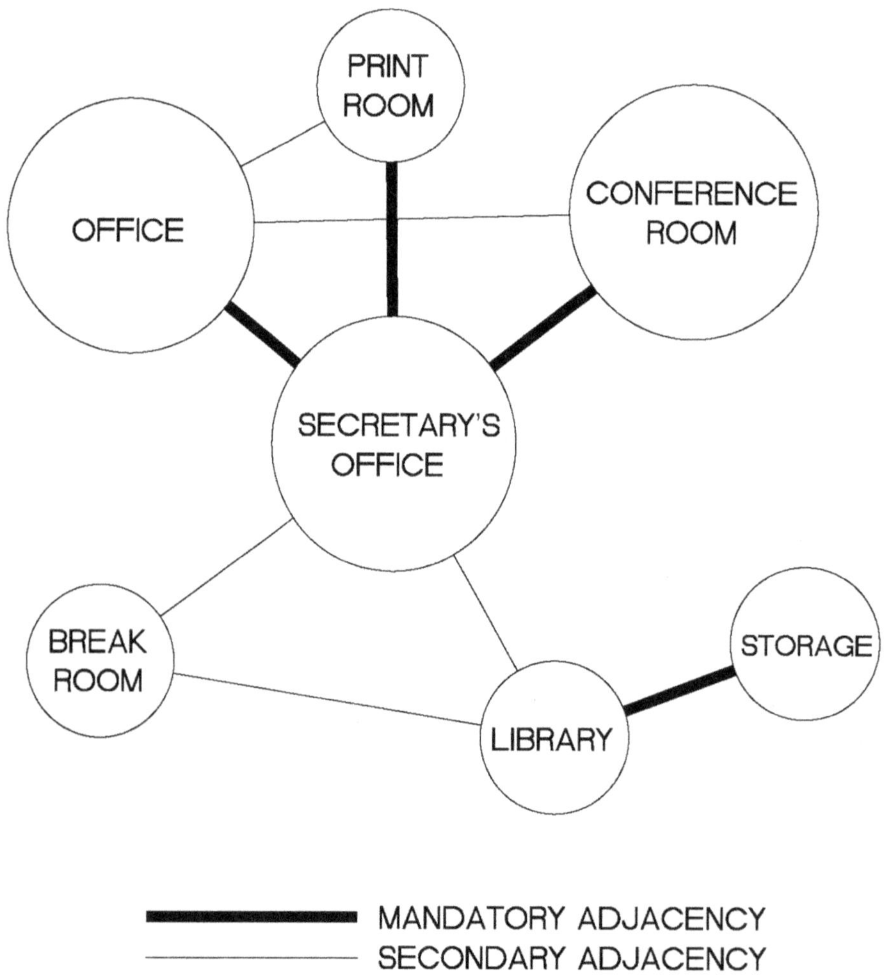

Figure 5.18 Adjacency bubble diagram

5. **Environmental considerations**
See previous discussions in Section B of this chapter.

6. **Views**
See previous discussions in Section A of Chapter Three.

7. **Vegetation diagrams**
An architect should be able to read landscape plans to coordinate the work of a landscape architect.

In landscape design firms, their main design and construction drawings are planting plans (planting design), and irrigation plans. (These irrigation plans are often done by consultants outside of landscape firms.)

1) **Planting presentation drawings**

For most of the landscape architect's offices, planting presentation drawings are only 10% to 20% of the workload. 80% to 90% of the planting design work is handled through planting plans in the construction drawing stage. Planting presentation drawings can be preliminary planting plans, except that they are normally color renderings, and shadows of plants are often shown to create a three-dimensional graphic effect (figure 5.19). The key for a successful landscape planting presentation drawing is to select good colors, especially a few good green colors, for plants. Each landscape office normally has a color palette for the colors that the office frequently uses for landscape planting presentation drawings.

2) **Planting plans**

For planting plans, graphic symbols and legends are normally simplified to communicate the design intent to the contractors in a most efficient way. There are two ways to organize and lay out planting plans:

a. **Type I layout (Figure 5.20)**

All plants are shown on the plans; the same kinds of plants are linked together with straight lines and then called out directly. A plant's botanical name, common name, and size and quantity are called out for each group of plants. If one plant has been used in different areas of a project, then this plant may be called out more than once in the same planting plan. Normally, the call-out line connects to the end of a string of plants, and the connecting line between plants is straight within the group and curved when crossing another connecting line. In the example shown in figure 5.20, some curved connecting lines are used within the group. These curved connecting lines are utilized because the layout of the planting plan is circular; using curved connecting lines will reduce the overlapping and crossing of various lines and make the planting plan easier to read. An exception has been made to the rules in the figure for clarity.

A plant list/schedule is also shown on the sheet, with the total number of each plant used shown on the list/schedule. This method was very commonly used in Landscape Architects' offices before **Computer Aided Design and Drafting (CADD)** became extensively used in Landscape Architecture and other design fields. The advantage of this method is that it is very straightforward: the name of the plant is right there next to the graphic symbols. It is easy to draw by hand: the graphic symbols do not need to be drawn exactly the same for the same plant, since they are linked together by straight lines. The disadvantage is that if the drawings need to be revised, often more than one location needs to be revised, since each plant has probably been called out several times in different areas of the project. If one or several groups of plants are missed in the revisions, conflicts and mistakes will occur. It is also more tedious and time-consuming to revise.

b. Type II layout

The other way is to use graphic symbols to represent plants: each kind of symbol represents a different kind of plant (figure 5.21). There is a planting legend that explains what each graphic symbol means (figure 5.22). Each graphic symbol is followed by the plant's botanical name/common name, size (size of the plant's container when it came from the plant nursery. This is the way nurseries size and sell their plants. It is widely used in the Landscape Architecture Practice), distance between plants (a designer should decide the proper distance between plants based on the final sizes of the plants), and quantities used. Plants are grouped as trees, shrubs, ground covers, lawn, vines, container, or pot planting, etc.

The advantage of this layout is that each plant is called out only once at the planting legend. If revisions are needed, a designer can keep the graphic symbol and simply change the plant name in the planting legend. The key is that the graphic symbols for the same plant need to look exactly the same at different areas of the project. The application of CADD has made this a very easy task. This type of layout is becoming very popular as the use of CADD is becoming more and more common. The biggest advantage of this layout is the saving of time in revisions, especially when the plants are changed, since you only need to change the plant call-out once at the planting legend. Anyone who has worked in a landscape architect's office would know revisions are very common for various reasons: owner's review comments, plan check corrections, availability of the plants specified, etc.

For both Type I and Type II planting plan layouts, common sizes for trees are 15 gallon (gal.), 24" box, 36" box, 48" box, and 60" box. Shrub sizes are often called out as 1 gal. at 24" On Center (O.C.), 5 gal. at 18" O.C., 5 gal. at 24" O.C., 5 gal. at 30" O.C., or 5 gal. at 36" O.C., etc. Ground covers are called out as 4" pot at 8" O.C., flats at 8" O.C., flats at 12" O.C., flats at 24" O.C., etc. Lawn is commonly called out as SOD (a very popular term for lawn in the U.S. Lawn is planted in the nurseries, and then harvested and cut out as 12" x 60" or 5 square foot flat pieces and then transported to project site and applied within a short time frame), and a designer customarily gives the total square footage of the lawn area for the project. Vines come in as 1 gal., 5 gal., and 4" pot sizes.

Soil areas that are not surfaced by ground cover or lawn are customarily covered by 1" diameter broken bark wood mulch to a depth of 2".

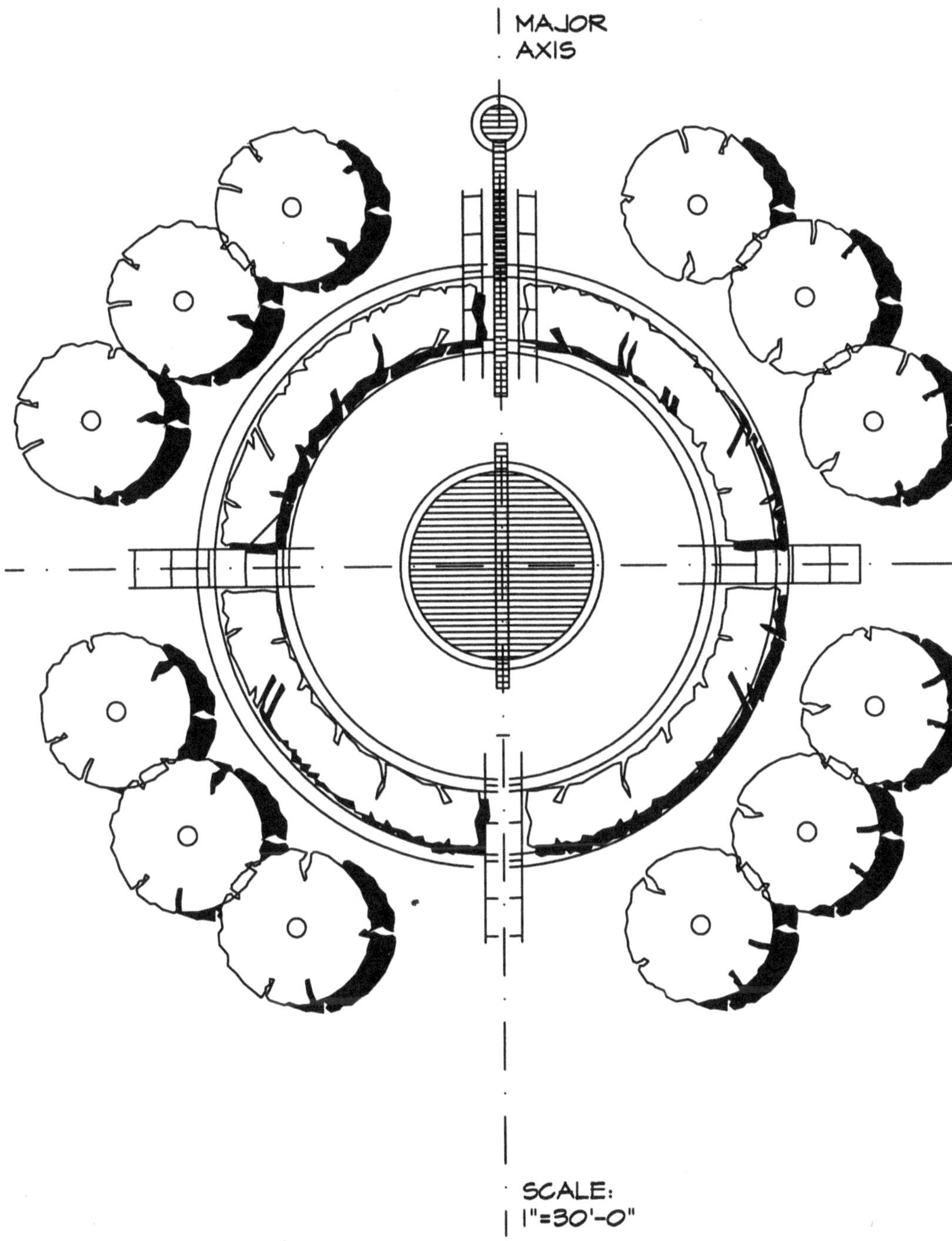

Figure 5.19 Planting presentation drawings

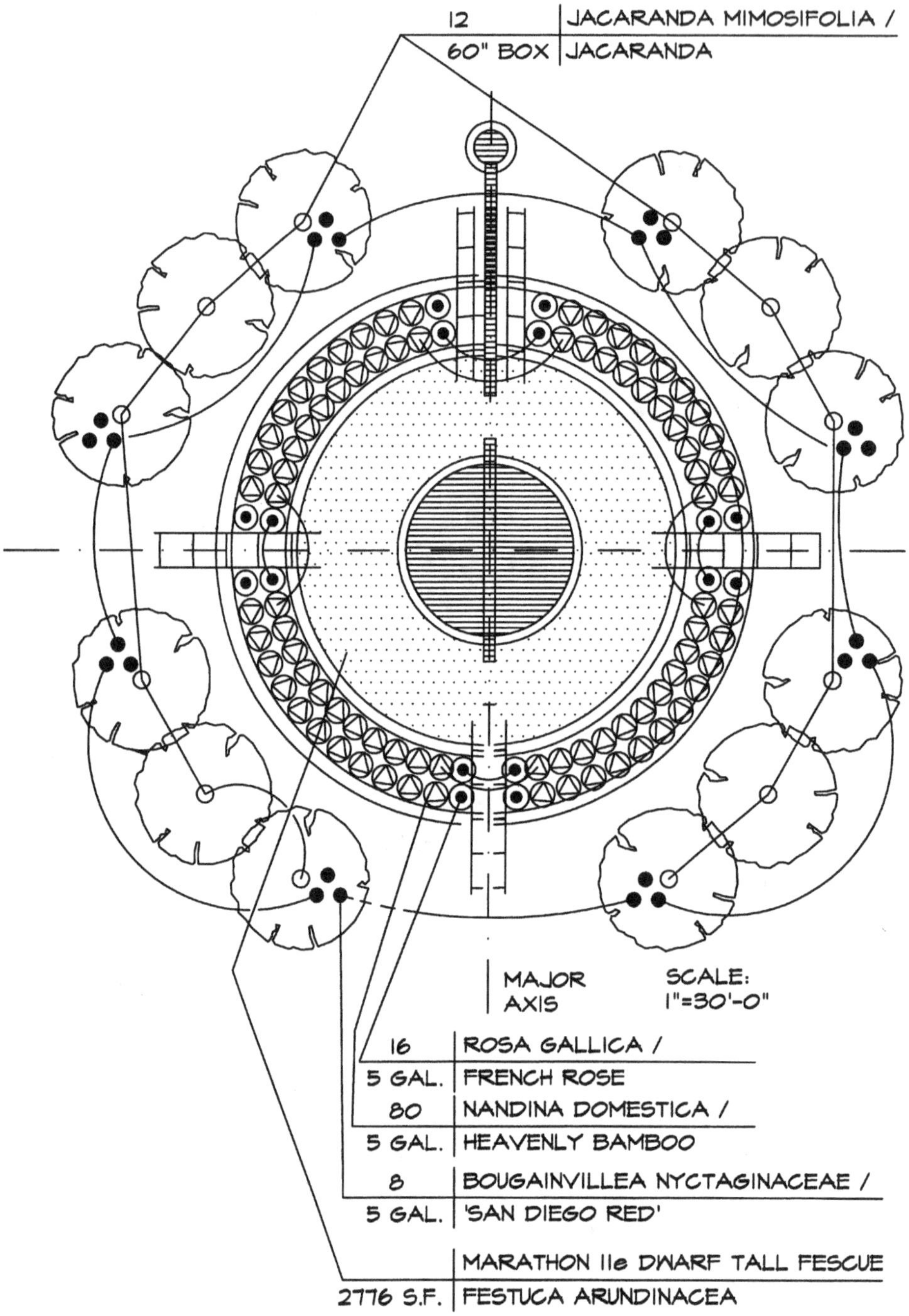

Figure 5.20 Type I layout for planting plan

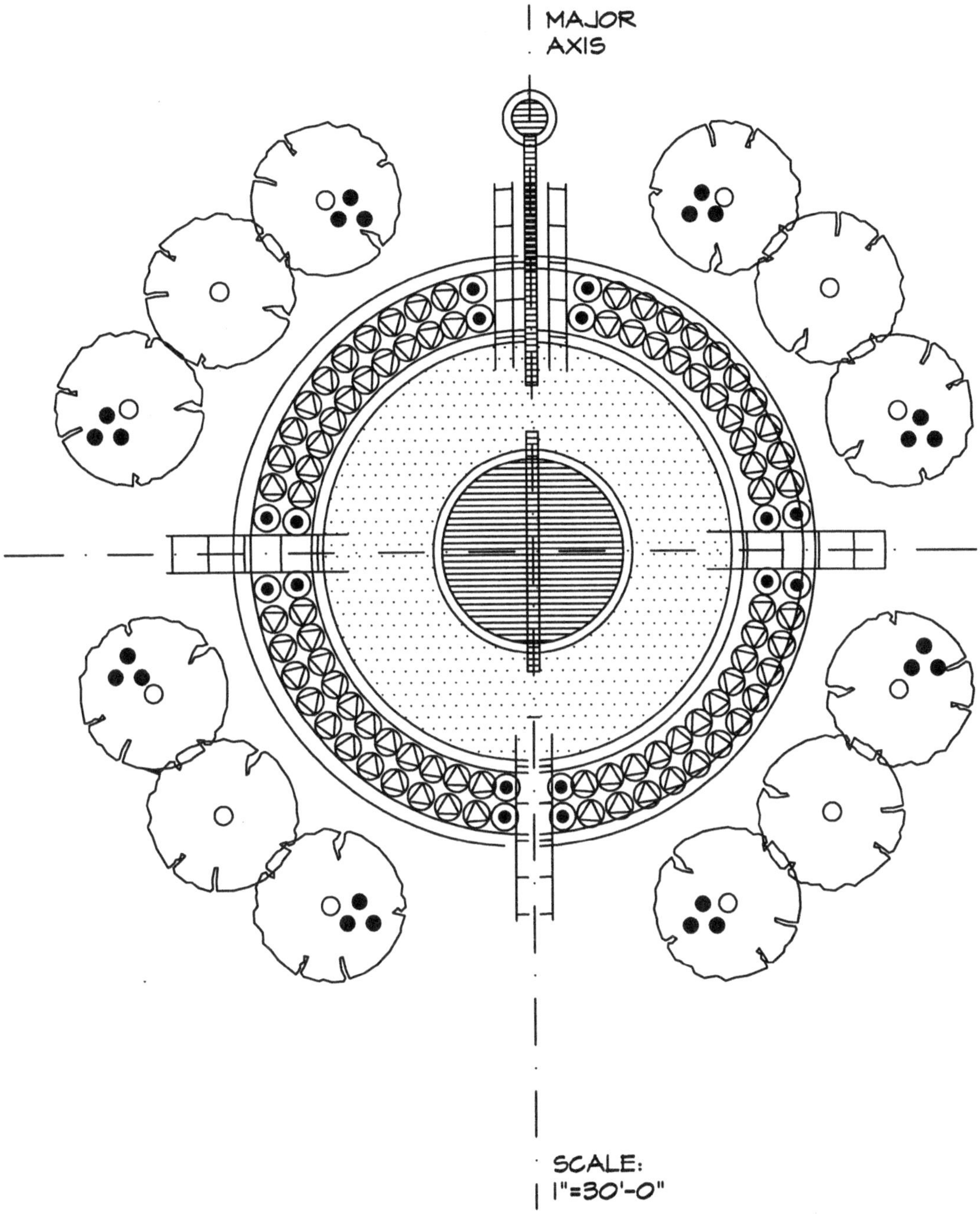

Figure 5.21 Type II layout for planting plan

PLANTING LEGEND

SYMBOL	BOTANICAL / COMMON NAME	SIZE	QUANTITY	DETAIL
TREES				
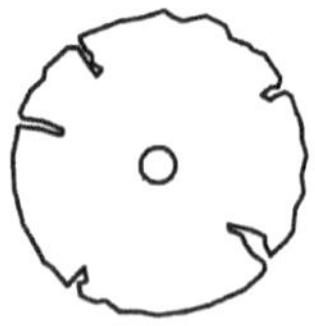	JACARANDA MIMOSIFOLIA / JACARANDA	60" BOX	12	-
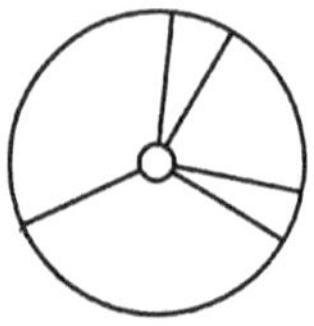	MAGNOLIA GRANDFLORA SOUTHERN MAGNOLIA	15 GAL.	48	-
	LIQUIDAMBER STYRACIFLUA AMERICAN SWEET GUM	36" BOX	18	-
SHRUBS				
	JASMINUM OLEACEAE / JASMINE	1 GAL. AT 18" O.C.	76	-
	NANDINA DOMESTICA / HEAVENLY BAMBOO	5 GAL. AT 36" O.C.	PER PLAN	-
	ROSA GALLICA / FRENCH ROSE	5 GAL. AT 24" O.C.	PER PLAN	-
GROUND COVERS				
	HEDERA HELIX / ENGLISH IVY	FLAT AT 12" O.C.	PER PLAN	-
	VINCA MINOR / DWARF PERIWINKLE	4" POT AT 8" O.C.	18	-
LAWN				
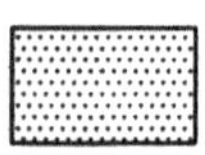	MARATHON IIe DWARF TALL FESCUE FESTUCA ARUNDINACEA		2776 S.F.	-
VINES				
	BOUGAINVILLEA NYCTAGINACEAE / 'SAN DIEGO RED'	5 GAL.	8	-

Figure 5.22 Sample planting legend

3) Other elements in landscape construction documents

We'll briefly introduce other factors in landscape construction documents here so that the less experienced designers or lay people can understand some basic elements of landscape practice.

Landscape construction documents are the legal documents prepared by the landscape architects. They are normally part of the binding contract documents between the owner and the contractor. **Landscape construction documents** include landscape construction drawings and landscape specifications. **Landscape specifications** cover quality and workmanship of landscape construction. It can be a separate specification book or can be set up as one or a few sheets within the construction drawing set. **Landscape construction drawings** cover location, size, quantity, and color of landscape elements. They are composed of planting plans (which we discussed in detail in this book), irrigation plans, some detail sheets, and hardscape plans if applicable. Irrigation plans are normally done by outside consultants in most landscape architects' offices. Important things to check in irrigation plans are:

a. **"Head to Head"** coverage: water from one irrigation sprinkler head shall always cover the adjacent irrigation sprinkler head to ensure even coverage (figure 5.23);
b. **"Zero run off":** make sure water from the sprinkler heads will not sprinkle onto or run off to adjacent properties;
c. **Water pressure:** Each sprinkler head shall have enough water pressure to operate at the design sprinkler radius;
d. **Oversizing the irrigation line:** it may be a good idea to oversize the irrigation line, because one may have to add more sprinkler heads and extension lines at a later addition to the original landscape project.

4) **Landscape details**

Landscape details are pretty much standard at each region, and most offices have a set of standard details to choose from to set up the detail sheets. Important things to consider are to coordinate the details with landscape planting plans and irrigation plans and make sure each detail is called out properly on the plans and modified if necessary per project conditions.

5) **Plant identification, plant list, and office library**

Most landscape architects and landscape offices have their most frequently used plant list and catalogs from nurseries set up in their office library. These lists and catalogs normally are adapted to the local conditions and can be very helpful for projects, especially for fast-track projects.

Plant materials are the vocabularies for planting design, and they are heavily affected by the local climate, weather, and soils conditions. There are plants commonly used in many different areas, but there are also many plants that are unique to a certain area. Even a very experienced landscape architect will have to learn plant materials and plant identification if he is relocated to a new area.

Plant materials are very important, but they are not the focus of this book:

First of all, this book focuses on an architect's ability to coordinate a landscape architect's work, and on the planting design principles and concepts, or the "grammar" and main ideas of planting design, which are the universal truth that can be used in different areas.

Figure 5.23 "Head to head" coverage in irrigation design

Second of all, many books on plant identification and on plant materials for different climate zones have already been published and are available. They can be very helpful for beginners in planting design or landscape architects who have been relocated to a new area.

Many colleges and universities also offer inexpensive courses on plant materials and plant identification. Students can take pictures of plants in class to assist them in their studies. Frequent visits to botanical gardens and nurseries are also effective ways to learn about plant materials and plant identification. The knowledge of plant materials and the skill of plant identification can only be obtained by each individual's long-term studies and accumulation. This is an area each individual needs to work on. It requires not only knowledge from books, but also actual field visits and practice.

6) Functional and ecological factors

Planting design often considers functional and ecological factors. **Functional requirements** generally come from the owner(s) and the users of the project. An architect can communicate with the owner(s) and the users and also visit the site to make personal observation to enlist all the functional needs. Many governing agencies have special landscape provisions. For example, some cities require a minimum distance between trees along the streets so as not to block view of motorists to ensure pedestrians' safety; other cities may require a minimum percentage of planting coverage for a project site or a minimum ratio between trees and the number of parking spaces. An architect may want to contact governing agencies (planning, public works, zoning, or building department, etc.) and obtain all relevant provisions at the beginning of the project and conform to them. This simple step can avoid major revisions at a later stage of planting design.

Ecology has been a major concern in the past decades. In planting design, this means you may choose plants that can adapt to the local climate and soil conditions and will not cause negative effects to existing local plant communities. This does *not* mean that designers are excluding the use of non-native plants. Throughout history, many civilizations have successfully used and domesticated non-native plants in gardens, and non-native plants have achieved great landscape effects in the U.S. and many other countries. Designers just need to choose plants more carefully and keep ecology in mind.

The **main ecological factors** include sunlight, air, water, soil, and temperature. Soil provides necessary "food" (minerals) for plants. The topsoil or surface soil often contains humus (organic decay) that can be very beneficial to the growth of plants. Many landscape architects would specify in their plan notes or specifications for the contractors to store and preserve the top 6" of surface soil or topsoil in the project site. The topsoil can be reapplied to the surface of the site after grading is finished. This is a good way to preserve and use the native topsoil of a site, which can contain many minerals and humus.

There are two common processes to improve soil: cultivation and fertilization.

Cultivation is a mechanical process that stirs the soil. It improves the aeration, drainage of soils, conserves moisture, and warms up soil. **Fertilization** increases the nutrients content and porosity of soil; it corrects pH value (alkalinity or acidity) and adds a specific mineral. Too much fertilizer can increase the acids in the soil and can be detrimental.

Mulches over soil surface can retain moisture and reduce the range of soil temperature changes. Sunlight, water, temperature, and wind also play important roles in plant life. Different plants have different needs for them.[3]

If you are interested in more information on landscape design and vegetation, you can refer to my other book, *Planting Design Illustrated*, it is available at:
http://www.GreenExamEducation.com

[3] Robinson, Florence Bell. *Planting Design.* Illinois: The Garrard Press, 1940. pp. 105-122.

Chapter Six

Building Analysis & Programming (37-43%)

NCARB Requirements
In this section, you'll analyze new or existing buildings relative to the program requirements, cost, and schedule. This is the largest section in the Programming & Analysis division.

A. Evaluate relevant qualitative and quantitative attributes of a new or existing building as they relate to the program (A/E)

NCARB Requirements
As an architect, you will need to review an existing or new building to determine if the design addresses views, daylight, spatial organization, and other program requirements. You will also need to analyze an existing building to determine if renovation, adaptive reuse, preservation, or demolition are appropriate and feasible to meet project needs. Relevant factors may include a building's Mechanical, Electrical, and Plumbing (MEP) systems, structural stability, hazardous materials, historic features, or eligibility for tax credits.

1. **Structural stability**
   A structural engineer consultant is typically needed for evaluating the structural stability, including foundation, slab, columns, beams, bearing walls, floor assemblies, roof assemblies, and overall building height, etc.

2. **Spatial organization**
   This involves evaluating the existing proposed spaces, and the possibility of reusing the existing spaces.

   The following are some basic forms of spatial organizations:
   - **Centralized organization** is several secondary spaces grouped around a dominant, central space which is typically large enough to accommodate several secondary spaces around its perimeter. The central organizing space may be either an exterior or interior space.
   - **Linear organization** contains a series of spaces that can either be linked through a separate and distinct linear space or be directly related to each other.
   - **Radial organization** includes a central space from which several organizations extend in a radial manner.
   - **Clustered organization** relies on physical proximity to create relationships among spaces. It often consists of spaces with similar functions, forms, or orientation.
   - **Grid organization** includes forms and spaces arranged in a three-dimensional grid pattern or field.

3. **MEP systems**
First, an architect is a specialist, and you need to have thorough and in-depth knowledge on architecture: you need to know architectural history, how to do sketch and perspective, and how to read and draw plans, elevations, sections, details, etc.

Second, an architect is a generalist or a "general contractor" for design service.

Many consultants, such as structural, electrical, mechanical, and plumbing engineers are design "subcontractors," and they subcontract a portion of the design work under the architect's master design contract.

Some other consultants such as soils, civil and fire protection engineers, and landscape architects are typically under the owner's contract. (They could be under the architect's contract also, depending on how the master design contract is structured.)

Regardless whose contract these consultants are under, as an architect, a generalist for design work, you do need to know enough about their portion of the work to coordinate with them.

For an existing building, an architect, with the assistance of mechanical, electrical, and plumbing engineers, need to evaluate the existing MEP systems to see if they can be adapted to the new use. For example, in a tilt-up concrete building, adding new HVAC units to the roof top may create extra weight that the existing building structure cannot handle, one possible solution is to place the heavy condensing units on the ground next to the building and use ductwork to route the air to various rooms.

Existing electrical and plumbing services should also be evaluated. For example, is the electrical panel sized adequately for the new uses? Are the existing sewer lines large enough? Is the number of plumbing fixtures adequate for the new use or occupancy?

4. **Views**
See previous discussions in Chapter Three.

5. **Solar exposure**
See previous discussions in Chapters Three, Four, and Five.

6. **Hazardous materials**
Hazardous materials include asbestos, lead-based paint, radon, formaldehyde, carbon monoxide, etc.

Asbestos is a mineral fiber that occurs in rock and soil.

Because of its fiber strength and heat resistance, asbestos has been used in a variety of building construction materials for insulation and as a fire retardant. Asbestos has also been used in a wide range of manufactured goods, mostly in building materials (roofing shingles, ceiling and floor tiles, paper products, and asbestos cement products), friction products (automobile clutch, brake, and transmission parts), heat-resistant fabrics, packaging, gaskets, and coatings.

One of EPA's priorities is to protect the public from adverse health effects of asbestos. Actions to protect the public from exposure to asbestos under the **Toxic Substances Control Act (TSCA)** include the following three pieces.

- **1989 Partial Ban on** the manufacture, import, processing, and distribution of some asbestos-containing products. EPA also banned new uses of asbestos which prevent new asbestos products from entering the marketplace after August 25, 1989. These uses remain banned. The April 2019 final rule does not provide a way for these uses to return to the marketplace.
- **April 2019 Final Rule to ensure that asbestos products that are no longer on the market cannot return to commerce** without the Agency evaluating them and putting in place any necessary restrictions or prohibiting use. The uses covered under this rule were not already prohibited under TSCA and could have returned to the market at any time.
- **Risk evaluation of asbestos under TSCA**. EPA is reviewing a handful of very limited, still ongoing uses of asbestos. The evaluation of the risks associated with ongoing uses of asbestos is required under TSCA section 6. If EPA finds unreasonable risk, the Agency will take prompt action to address those risks.

Where asbestos may be found:

- Attic and wall insulation produced containing vermiculite
- Vinyl floor tiles and the backing on vinyl sheet flooring and adhesives
- Roofing and siding shingles
- Textured paint and patching compounds used on walls and ceilings
- Walls and floors around wood-burning stoves protected with asbestos paper, millboard, or cement sheets
- Hot water and steam pipes coated with asbestos material or covered with an asbestos blanket or tape
- Oil and coal furnaces and door gaskets with asbestos insulation
- Heat-resistant fabrics
- Automobile clutches and brakes

Asbestos fibers may be released into the air by the disturbance of asbestos-containing material during product use, demolition work, building or home maintenance, repair, and remodeling. In general, exposure may occur only when the asbestos-containing material is disturbed or damaged in some way to release particles and fibers into the air.

Exposure to asbestos increases your risk of developing lung disease. That risk is made worse by smoking. In general, the greater the exposure to asbestos, the greater the chance of developing harmful health effects.

Disease symptoms may take many years to develop following exposure.

Asbestos-related conditions can be difficult to identify. Healthcare providers usually identify the possibility of asbestos exposure and related health conditions like lung disease by taking a thorough medical history. This includes looking at the person's medical, work, cultural and environmental history.

After a doctor suspects an asbestos-related health condition, he or she can use a number of tools to help make the actual diagnosis. Some of these tools are physical examination, chest x-ray and pulmonary function tests. Your doctor may also refer you to a specialist who treats diseases caused by asbestos.

Three of the major health effects associated with asbestos exposure are:

- lung cancer
- mesothelioma, a rare form of cancer that is found in the thin lining of the lung, chest and the abdomen and heart
- asbestosis, a serious progressive, long-term, non-cancer disease of the lungs

See link below for more information:
https://www.epa.gov/asbestos/learn-about-asbestos#asbestos

Lead is a naturally occurring element found in small amounts in the Earth's crust. While it has some beneficial uses, it can be toxic to humans and animals, causing health effects.

Lead can be found in all parts of our environment – the air, the soil, the water, and even inside our homes. Much of our exposure comes from human activities including the use of fossil fuels including past use of leaded gasoline, some types of industrial facilities, and past use of lead-based paint in homes. Lead and lead compounds have been used in a wide variety of products found in and around our homes, including paint, ceramics, pipes and plumbing materials, solders, gasoline, batteries, ammunition, and cosmetics.

Lead may enter the environment from these past and current uses. Lead can also be emitted into the environment from industrial sources and contaminated sites, such as former lead smelters. While natural levels of lead in soil range between 50 and 400 parts per million, mining, smelting, and refining activities have resulted in substantial increases in lead levels in the environment, especially near mining and smelting sites.

When lead is released to the air from industrial sources or spark-ignition engine aircraft, it may travel long distances before settling to the ground, where it usually sticks to soil particles. Lead may move from soil into ground water depending on the type of lead compound and the characteristics of the soil.

Federal and state regulatory standards have helped to reduce the amount of lead in air, drinking water, soil, consumer products, food, and occupational settings.

If a home was built before 1978, there is a good chance it has lead-based paint. In 1978, the federal government banned consumer uses of lead-containing paint, but some states banned it even earlier. Lead from paint, including lead-contaminated dust, is one of the most common causes of lead poisoning.

- Lead paint is still present in millions of homes, sometimes under layers of newer paint. If the paint is in good shape, the lead paint is usually not a problem. Deteriorating lead-based paint (peeling, chipping, chalking, cracking, damaged, or damp) is a hazard and needs immediate attention.
- It may also be a hazard when found on surfaces that children can chew or that get a lot of wear-and-tear, such as:
  - Windows and windowsills;
  - Doors and door frames; and
  - Stairs, railings, banisters, and porches.

- Be sure to keep all paint in excellent shape and clean up dust frequently. Read about simple steps to protect your family from lead hazards (PDF): https://www.epa.gov/lead/protect-your-family-lead-your-home

- Lead in household dust results from indoor sources such as deteriorating lead-based paint.
- Lead dust can also be tracked into the home from soil outside that is contaminated by deteriorated exterior lead-based paint and other lead sources, such as industrial pollution and past use of leaded gasoline. Read more about lead dust: https://www.epa.gov/lead/protect-your-family-exposures-lead#sources

- Renovation, repair or painting activities can create toxic lead dust when painted surfaces are disturbed or demolished. Learn more about hiring lead-safe certified contractors. https://www.epa.gov/lead/renovation-repair-and-painting-program-consumers

- Pipes and solder - Lead is used in some water service lines and household plumbing materials. Lead can leach, or enter the water, as water flows through the plumbing. Lead pipes and lead solder were commonly used until 1986. Read more about lead in drinking water: https://www.epa.gov/lead/protect-your-family-exposures-lead#water

Volatile organic compounds (VOCs) are emitted as gases from certain solids or liquids. VOCs include a variety of chemicals, some of which may have short- and long-term adverse health effects. Concentrations of many VOCs are consistently higher indoors (up to ten times higher) than outdoors. VOCs are emitted by a wide array of products numbering in the thousands.

Organic chemicals are widely used as ingredients in household products. Paints, varnishes and wax all contain organic solvents, as do many cleaning, disinfecting, cosmetic, degreasing and hobby products. Fuels are made up of organic chemicals. All of these products can release organic compounds while you are using them, and, to some degree, when they are stored.

EPA's Office of Research and Development's **"Total Exposure Assessment Methodology (TEAM) Study"** (Volumes I through IV, completed in 1985) found levels of about a dozen common organic pollutants to be 2 to 5 times higher inside homes than outside, regardless of whether the homes were located in rural or highly industrial areas. TEAM studies indicated that while people are using products containing organic chemicals, they can expose themselves and others to very high pollutant levels, and elevated concentrations can persist in the air long after the activity is completed.

Sources of VOCs:
Household products, including:
- paints, paint strippers and other solvents
- wood preservatives
- aerosol sprays
- cleansers and disinfectants
- moth repellents and air fresheners
- stored fuels and automotive products
- hobby supplies
- dry-cleaned clothing
- pesticide

Other products, including:
- building materials and furnishings
- office equipment such as copiers and printers, correction fluids and carbonless copy paper
- graphics and craft materials including glues and adhesives, permanent markers and photographic solutions.

Reducing VOCs can help a project team **Improve Indoor Air Quality (IAQ)** and gain points in the LEED building rating system.

See link below for more information:
https://www.epa.gov/indoor-air-quality-iaq/volatile-organic-compounds-impact-indoor-air-quality

Formaldehyde is a colorless, flammable gas at room temperature and has a strong odor. Exposure to formaldehyde may cause adverse health effects.

Formaldehyde is found in:
- Resins used in the manufacture of composite wood products (i.e., hardwood plywood, particleboard and medium-density fiberboard);
- Building materials and insulation;
- Household products such as glues, permanent press fabrics, paints and coatings, lacquers and finishes, and paper products;

- Preservatives used in some medicines, cosmetics, and other consumer products such as dishwashing liquids and fabric softeners; and
- Fertilizers and pesticides.

It is a byproduct of combustion and certain other natural processes, and so is also found in:

- Emissions from un-vented, fuel burning appliances, like gas stoves or kerosene space heaters; and
- Cigarette smoke.
- Learn more from the Consumer Product Safety Commission about products that contain formaldehyde.

The primary way people can be exposed to formaldehyde is by breathing air containing off-gassed formaldehyde. Everyone is exposed to small amounts of formaldehyde in the air that has off-gassed from products, including composite wood products.

Formaldehyde can cause irritation of the skin, eyes, nose, and throat. High levels of exposure may cause some types of cancers.

See the following link for more information:
https://www.epa.gov/formaldehyde

Carbon Monoxide (CO) is an odorless, colorless and tasteless gas. It is highly toxic to humans and animals in higher quantities.

Because it is impossible to see, taste or smell, breathing CO can incapacitate or kill you before you are aware it is present in your home. At lower concentrations, CO may produce flu-like symptoms, including headache, dizziness, disorientation, nausea and fatigue.

Sources of CO include:

- unvented kerosene and gas space heaters
- leaking chimneys and furnaces
- back-drafting from furnaces, gas water heaters, wood stoves and fireplaces
- gas stoves
- generators and other gasoline powered equipment
- automobile exhaust from attached garages
- tobacco smoke
- auto, truck, or bus exhaust from attached garages, nearby roads, or parking areas
- incomplete oxidation during combustion in gas ranges, and unvented gas or kerosene heaters
- worn or poorly adjusted and maintained combustion devices (e.g., boilers, furnaces)
    - if the flue is improperly sized, blocked or disconnected
    - if the flue is leaking

Health Effects Associated with Carbon Monoxide include:

At low concentrations:
- fatigue in healthy people
- chest pain in people with heart disease

At moderate concentrations:
- angina
- impaired vision
- reduced brain function

At higher concentrations:
- impaired vision and coordination
- headaches
- dizziness
- confusion
- nausea
- flu-like symptoms that clear up after leaving home
- fatal at very high concentrations

Acute effects are due to the formation of carboxyhemoglobin in the blood, which inhibits oxygen intake.

See the following link for more information:
https://www.epa.gov/indoor-air-quality-iaq/carbon-monoxides-impact-indoor-air-quality

See Chapter Three for more discussions on radon and lead.

7. **Other attributes**
Other attributes include the building styles, context, materials, and colors, etc.

B. Evaluate documentation, reports, assessments, and analyses to inform the building program (A/E)

NCARB Requirements
In addition to analyzing the attributes of a building, you will need to review and interpret technical documentation such as structural assessments, geotechnical reports, neighborhood and regional context, and condition assessment reports. These are used to consider the feasibility of a project and how the building program may be positively or negatively affected.

1. **Geotechnical reports**
See discussion in Chapter Five.

2. **Structural assessments**
See previous discussion in this chapter.

3. **Utility surveys**
   See discussion in Chapter Five. Utility surveys can be part of the site survey, and the utility locations can be marked on the survey map.
4. **Topographic maps**
   See discussion in Chapter Five.

5. **Demographics**
   See discussion in Chapter Five.

6. **Regional context**
   Regional context in a particular region is an architectural style developed and based on the history, social influence, climate, geology, geography, and topography of the region.

   Contextual architecture, or **contextualism** is a philosophical approach to design a structure in response to the abstract and literal characteristics of its environment.

   Within contextual architecture there are three categories: **vernacular architecture**, **regional architecture**, and **critical regionalism**.

   Vernacular architecture uses local materials and knowledge, and represents the majority of buildings and settlements created in pre-industrial societies which encompassed 95% of the world's built environment in 1969. It includes a very wide range of buildings, traditions, and methods of construction.

   Regional architecture closely follows the developments of vernacular architecture but uses building technologies and materials.

   Critical regionalism strives to counter the lack of identity and sense of place (**genius loci**) of the International Style, but also rejects the fanciful ornamentation and individualism of Postmodern architecture. Its stylings seek to provide an architecture rooted in the modern tradition, yet tied to cultural geographical context. Critical regionalism is not simply regionalism in the sense of vernacular architecture. This approach seeks to balance between the global and the local languages of architecture.

   Note:
   In recent years, the Latin word, ***"genius loci"*** *(pronounced as "loh-kai") has become very popular in architecture education and practice. This term means the particular spirit of a site and is a belief that each site has a special importance, a particular meaning beyond function and beauty. This approach can be a very good way to start the fundamental concept for the design of a project. You can do various research works on the site's history and context, try to define the "genius loci" of the site, and base your design on this interpretation. This is a general philosophical concept that came from ancient Western civilizations. This approach focuses on the study of the objective site.*

7. **Environmental data**
   Besides our discussion in previous chapters, the EPA provides an environmental dataset gateway and features data products at the following link:
   https://edg.epa.gov/metadata/catalog/main/home.page
8. **Historic reports**
   See discussion in Chapters Four and Five.

9. **Other similar information**
   Other site related reports include title reports, natural hazard disclosure reports, feasibilities studies, financial reports, etc.

C. Identify and prioritize components of the building program (A/E)

NCARB Requirements
As an architect, you must be able to review and understand the building program relative to the client's requirements. This includes the consideration of primary vs. subsidiary, back of house vs. front of house, occupied vs. unoccupied, as well as phasing and major circulation components. Understanding the area requirements of a building type relative to the program, including net square footage and gross square footage, is very important, as are how they relate to each other and the overall program requirements.

1. **Prioritizing components of the building program**
   Architecture is a balancing art. An architect needs to set up an order of importance amongst different project requirements, like the position of spaces, sizes of spaces, and values or goals of the client. Consideration should be given to whether a space is primary or subsidiary, should be placed at back or front of a building, and is occupied or unoccupied.

   If a client emphasizes the corporation's image, maybe the reception area and conference as well as the building entrance need to have high priority for that individual office.

2. **Phasing and major circulation components**
   If a large project cannot be built all at once, then phasing is a good option. In fact, most of residential development is done in several phases. The developer will build a few typical model homes for the release of phase one. After phase one is completed, the developer makes some profit and can use a portion of that money to build the subsequent phases. Phasing is a good way to limit the risk of investment too, since the investment for building one phase is a fraction of the budget of building the entire project at once.

   Major circulation components for a building are also important. For example, for a commercial building, the most important circulation route is the customer route, and other routes such as delivery or maintenance need to be secondary. For a hospital, it is important to separate the contagious diseases area from the regular patients to avoid cross-infection.

3. Net square footage and gross square footage

Per National Center for Educational Statistics (NCES), **Net Square Footage** or Assignable Area (Net Assignable Square Feet - NASF) is the sum of all areas on all floors of a building assigned to, or available for assignment to, an occupant or specific use.

Net Assignable Area is computed by physically measuring or scaling measurements from the inside faces of surfaces that form the boundaries of the designated areas. Exclude areas having less than a 3-foot clear ceiling height unless the criteria of a separate structure are met. See figure 6.1.

Gross Square Feet (GSF) or Gross Area is the sum of all areas on all floors of a building included within the outside faces of its exterior walls, including all vertical penetration areas, for circulation and shaft areas that connect one floor to another.

Gross Area is computed by physically measuring or scaling measurements from the outside faces of exterior walls, disregarding cornices, pilasters, buttresses, etc., that extend beyond the wall faces. Exclude areas having less than a 3-foot clear ceiling height unless the criteria of a separate structure are met. See figure 6.2.

Measured in terms of gross square feet (GSF):
GSF = net assignable area + tare are

Note:
***Tare area** is the remainder after the net assignable area is subtracted from the gross building area.*

Efficiency of the building or net-to-gross ratio or **efficiency ratio** is the net square footage divided by the gross square feet.

$$\textbf{efficiency of the building} = \frac{\textbf{net assignable area}}{\textbf{gross square feet}}$$

An architect may need to help a tenant to determine the **rentable space**. The rentable space is the **usable space** (or occupant area) plus circulation and services *within* the tenant space for the tenant.

A **load factor** equals the rentable space divided by usable space:

$$\textbf{load factor} = \frac{\textbf{rentable space}}{\textbf{usable space}}$$

Note:
Efficiency ratio deals with the efficiency of the entire building, and the load factor deals with the efficiency of a tenant's space. You should become familiar with the two formulas above. I'd not be surprised if NCARB tests your ability to use both of the formulas in the same question.

Building Owner and Managers Association International (**BOMA International**) and the International Facility Management Association (**IFMA**) both publish standards on how to measure a building. It is up to the owner to choose which standard to use.

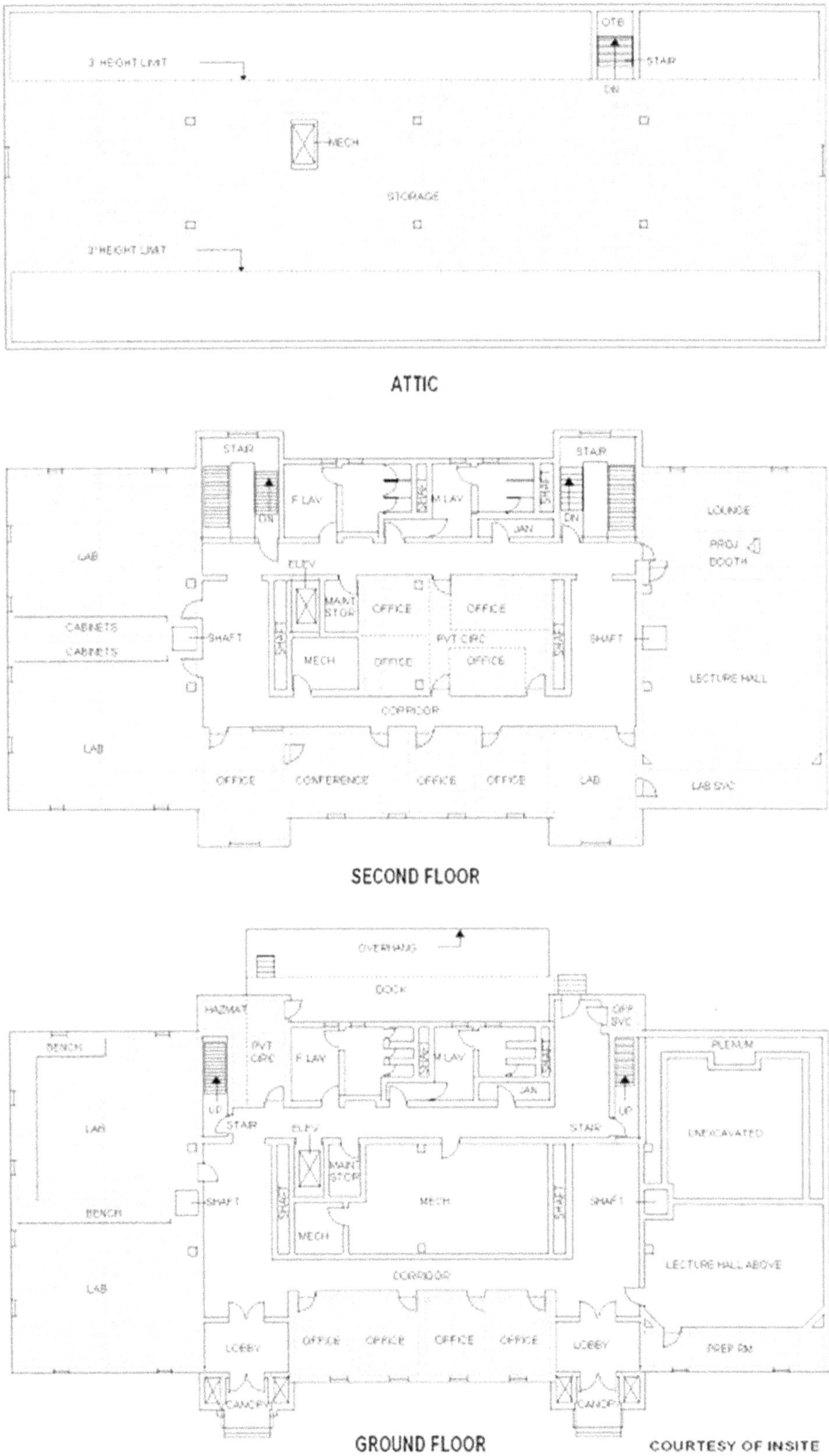

Figure 6.1 Net Assignable Area of a building by floor (Source: NCES)

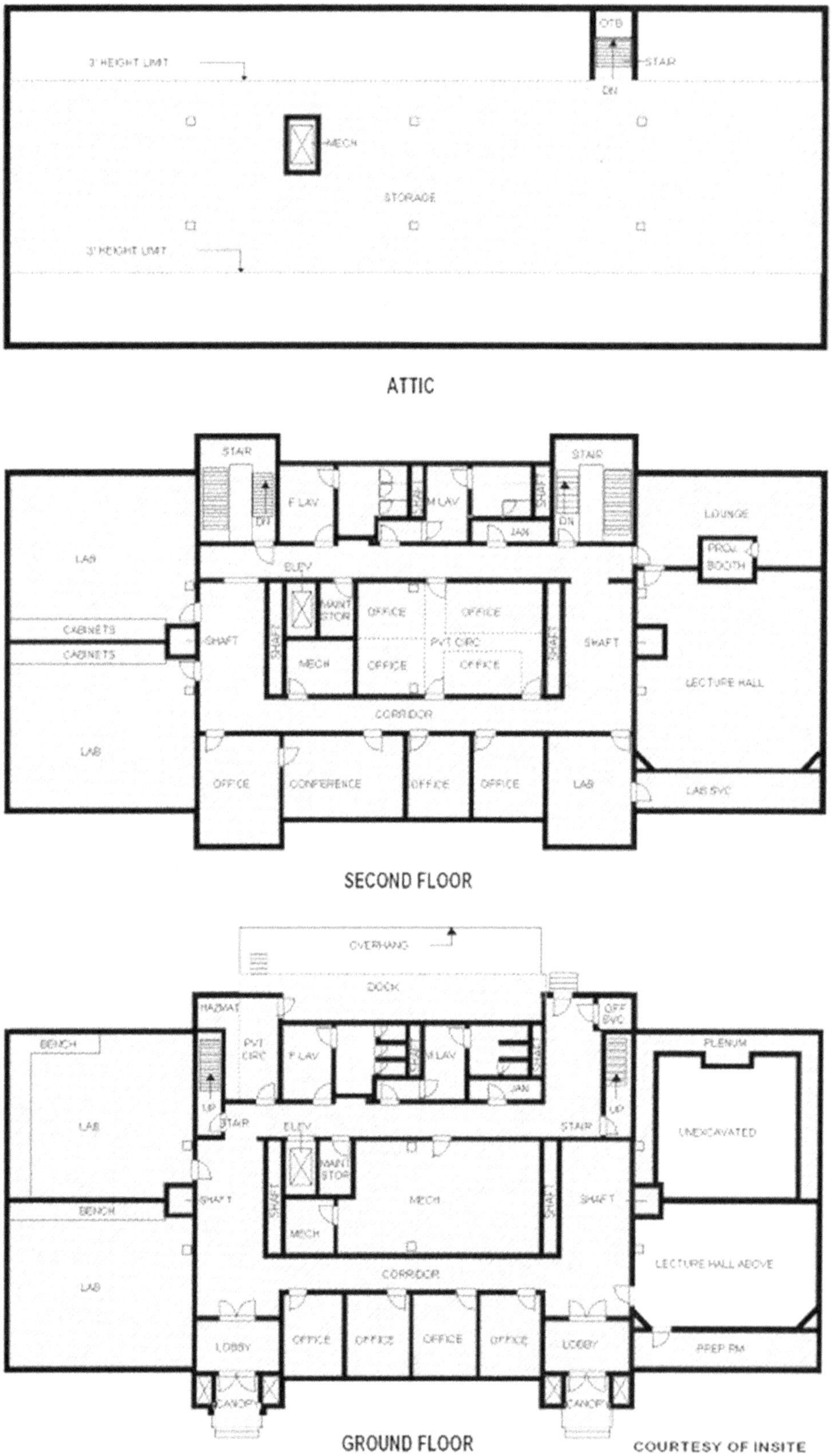

Figure 6.2 Gross Area of a building by floor (Source: NCES)

D. Assess spatial and functional relationships for the building program (A/E)

NCARB Requirements
Spatial organization is an important aspect of architecture, so reviewing and analyzing the building program relative to both horizontal and vertical spatial and functional relationships is an important assessment. You must consider the vertical relationships of shafts, stairs, conveying systems, atriums, and other multi-level spaces, as well as the horizontal relationships including circulation, entry, loading, assembly, MEP spaces, and other programmed spaces.

You can use the adjacency matrix and adjacency diagram (or bubble diagram) to assist you in the spatial organization. See discussions in Chapter Five.

1. **Horizontal spatial relationships**
   Entry, loading, and rooms with heavy MEP equipment are normally located at the ground level. Each floor has circulation requirements that can be met by corridors or intervening spaces.

2. **Vertical spatial relationships**
   Some spaces such as shafts, stairs, conveying systems, and atriums may span multiple levels, and require extra coordination efforts. For example, an escalator has an important working point at each level per the manufacturer's requirement. Critical information such as this should be coordinated with the manufacturer as early as possible.

3. **Functional relationships**
   Some spatial relationships are mandated by the function of the building. Some spaces may need to be adjacent to each other and on the same floor. Some spaces may need to be accessible to each other. The program may also require some spaces be placed at the same level. All these considerations should be balanced and considered at the early stage of a project.

E. Recommend a preliminary project budget and schedule (U/A)

NCARB Requirements
As an architect, you must be able to compare available project information (master plans, existing documentation, program requirements, site information, etc.) to determine a preliminary project schedule and cost. If the project budget and schedule have been defined by the client, you should be able to evaluate this information to determine the feasibility of the project and provide appropriate recommendations.

The project costs include the costs for building, site development, furniture, fixtures and equipment (FF&E), professional services, and financing, etc. The four key elements of a budget are quantity, quality, time, and available funds.

1. **Ways to establish or evaluate a project budget**
   There are several ways to establish or evaluate a project budget:

   1) **A pro forma analysis**
      A pro forma analysis is a set of calculations that estimate the profits real estate development is likely to create. To make the project feasible and profitable, a limit is set on the project budget.

   2) **Project comparison method using historical data**
      Use the cost of a previously built and similar project to establish the budget. You need to add additional cost to cover inflation over the years, and a contingency amount.

   3) **Unit price method**
      It uses average unit cost such as cost per square foot (**area method** or square-foot method), or cost per cubic feet (**volume method**) to establish the budget.

   4) **Assembly method or system method**
      It uses the cost of each type of major subsystems to calculate the costs and add them up to reach the overall project costs.

2. **A project's schedule**
   Your client may request you to provide a project schedule at the beginning of the project or at any other stage of the project. For you, it is easier to come up with a project schedule at the design development stage because you already know the materials, methods, major building systems (including structural and mechanical systems), and construction type and standards for your project. You should have a clear understanding of the project timeframe by now.

   As long as you know how much time you need for each phase of the project, you can easily come up with a project schedule with **Microsoft Project** software.

   For example, the following are the major phases and activities of a retail project for each phase and some estimate times to assist you to come up with a project schedule. Not all steps and phases are necessary. Contact your governing agencies to confirm.

   1) **Entitlement and design**

      a. Produce a design submittal package for the city, including overall building elevations, preliminary site plans, and landscape plans. It normally takes 3 to 4 weeks. Verify with the city regarding the exact requirements for the design submittal package because its content will affect the schedule.
      b. The client's review of your design submittal package normally takes 2 to 3 weeks.

c. The city planning department staff will review your design submittal package. It normally takes 2 to 3 weeks. Please verify with the city regarding the exact turnaround time.
d. Review by Design Review Board (DRB) or Architectural Review Committee (ARC). You normally need to go through this step before you can present to the planning commission. You need to contact the planning department to find out the schedule and date for DRB or ARC meetings and update your schedule accordingly.
e. Review by the city's planning commission. You need to contact the planning department to find out the schedule and date for planning commission meetings and update your schedule accordingly.
f. Review by the city council. You need to contact the planning department to find out the schedule and date for city council meetings and update your schedule accordingly.
g. Design appeal period. It normally takes 2 to 3 weeks. You need to contact the planning department to confirm and update your schedule accordingly.

2) Design development

a. It normally takes 3 to 4 weeks.
b. This step can be omitted.

3) Construction documents

a. Site development construction documents

a) Coordinate with the civil engineer to come up with a base site plan for site architectural, landscape, and electrical plans. It normally takes 2 to 3 weeks.
b) Produce site construction documents, including your plans and the consultants' plans. It normally takes 4 to 6 weeks. You need to contact the consultants to confirm and update your schedule accordingly.
c) Review by governing agencies and clients. They can occur concurrently and normally take 2 to 3 weeks. You need to contact governing agencies to confirm and update your schedule accordingly.

Note:
Civil plans are normally submitted separately by the civil engineer to the public works department.

d) Update plans per plan check correction list and client's comments. It normally takes 2 to 3 weeks.
e) The 2nd plan check by governing agencies. It normally takes 2 to 3 weeks.
f) Permit ready.

b. Building construction documents

This can be concurrent with the site development construction documents.

a) Coordinate with your client and consultants to come up with the base plans for building architectural, structural, electrical plans, mechanical, and plumbing plans. It normally takes 2 to 3 weeks.
b) Produce building construction documents, including your plans and the consultants' plans. It normally takes 6 to 8 weeks. You need to contact the consultants to confirm and update your schedule accordingly.
c) Review by governing agencies and clients. They can occur concurrently and normally take 2 to 3 weeks. You need to contact governing agencies to confirm and update your schedule accordingly.
d) Update plans per plan check correction list and client's comments. It normally takes 2 to 3 weeks.
e) The 2nd plan check by governing agencies. It normally takes 2 to 3 weeks.
f) Permit ready.

4) **Bidding and negotiation**

a. Bidding normally takes 3 to 4 weeks.
b. Negotiation and awarding the contract normally takes 1 week.

5) **Site and building construction**

a. It normally takes 3 to 4 weeks to get a certified pad. This need to happen before the building permit is ready.
b. Construction normally takes 26 to 28 weeks for retail buildings. It can take longer for other building types.

F. Identify alternatives for building and structural systems for given programmatic requirements, preliminary budget, and schedule (U/A)

NCARB Requirements
You will need to understand various building materials as well as civil, structural, mechanical, electrical, and plumbing systems to determine the appropriateness of each to meet project needs, based on factors such as building type, function, program, availability, cost, and sustainability goals. Even before the selection of each system, you will need to be able to identify alternative systems that could fulfill the project requirements.

1. Architectural systems

Based on building type, function, program, availability, schedule, and cost, an architect needs to be able to select the proper architectural systems. For example, a building may need to use brick veneer instead of regular plaster because of the context of the project. A warehouse type wholesale store like Costco is better served with an exposed ceiling. An office may need a regular T-bar ceiling with 2x4 or 2x2 ceiling grid. The neighborhood or HOA requirements may require a building to have tile roof instead of composition or wood shingle roof, etc.

2. **Civil systems**

Civil systems include storm drains, open drainage systems, and site retaining walls, etc.

Because of the green building trend, more and more projects are avoiding the excessive use of traditional storm drains, and try to use perforated concrete, grass-covered open swales, wetlands, detention basins, and retention basins, etc. to increase re-charging of the ground water, reuse rainwater, and avoid run-offs.

A **detention basin** or retarding basin is an excavated area adjacent to rivers, streams, lakes or bays to protect against flooding, and downstream erosion by storing water for a limited time. These basins are also called "holding ponds", "**dry ponds**" or "dry detention basins" if no *permanent* pool of water exists.

Detention ponds that are designed to *permanently* retain some volume of water at all times are called **retention basins**. In its basic form, a detention basin is used to manage water quantity while having a limited effectiveness in protecting water quality, unless it includes a permanent pool feature.

A **retention basin**, wet pond, wet detention basin, or stormwater management pond is an artificial pond with vegetation around the perimeter and includes a *permanent* pool of water. The area is used to prevent flooding and downstream erosion, improve water quality in adjacent rivers, streams, lakes or bays, and manage stormwater runoff.

The **Clean Water Act (CWA)** prohibits anybody from discharging "pollutants" through a "point source" into the "water of the United States" unless they have a **National Pollutant Discharge Elimination System (NPDES)** permit. The permit will contain limits on what you can discharge, monitoring and reporting requirements, and other provisions to ensure that the discharge does not hurt water quality or public health. In essence, the permit translates general requirements of the Clean Water Act into specific provisions tailored to the operations of each person discharging pollutants.

See the following link for more information:
https://www.epa.gov/npdes

Best management practices (BMPs) are types of water pollution control in the United States and Canada. Historically the term has referred to auxiliary pollution controls in the fields of municipal sewage control and industrial wastewater control. In wetland management and stormwater management (both urban and rural), BMPs may refer to a treatment technique or principal control.

Congress referred to BMP in several sections of the U.S. Clean Water Act (CWA) but did not define the term.

In implementing the CWA, the U.S. Environmental Protection Agency (EPA) defined BMP in the federal wastewater permit regulations, initially to refer to auxiliary procedures for industrial wastewater controls as "schedules of activities, prohibitions of practices, maintenance procedures, and other management practices to prevent or reduce the pollution of waters of the United States. BMPs also include treatment requirements, operating procedures, and practices to control plant site runoff, spillage or leaks, sludge or waste disposal, or drainage from raw material storage."

Later EPA added a reference to stormwater management BMPs stating, "each NPDES permit shall include conditions meeting the following requirements when applicable... (1) Best management practices (BMPs) to control or abate the discharge of pollutants when... (2) Authorized under section 402(p) of the CWA for the control of storm water discharges..."

3. **Structural systems**
   1) **Standard structural systems**
      a. **Wood** is inexpensive, plentiful, and relatively strong in both tension and compression. It is typically used in **one-way structural systems.**

         a) **Joists** are horizontal members of support. Typical size are 2x6, 2x8, 2x10, and 2x12. There are nominal dimensions. See figure 6.3 for their actual dimensions. Typical joist spacing is 12". 16", or 24" on center. Typical spans of the joists are between 20' and 25'.

            Plywood, oriented strand board (**OSB**), or particleboard subflooring spans the space between the joists.

         b) **Wood beams** with nominal widths of 4" or 6" are used to span between the bearing walls and girders at spacing of 4', 6' or 8'. Laminated or solid wood decking span between the wood beams. The span of wood beams are between 10' and 20'.

Lumber Dimension	
Nominal	**Actual**
2 x 2	1-1/2" x 1-1/2"
2 x 4	1-1/2" x 3-1/2"
2 x 6	1-1/2" x 5-1/2"
2 x 8	1-1/2" x 7-1/4"
2 x 10	1-1/2" x 9-1/4"
2 x 12	1-1/2" x 11-1/4"
3 x 6	2-1/2" x 5-1/2"
4 x 4	3-1/2" x 3-1/2"
4 x 6	3-1/2" x 5-1/2"

Figure 6.3 Lumber dimension

         c) **Glulam (glued-laminated) construction** uses structural members that are made up of lumber pieces, ¾" or 1 ½" thick, that are glued together in a factory. The standard widths of glulam are 3 1/8", 5 1/8", 6 ¾" and 8 ¾". Glulam can span between 15' to 60'.

d) **Lightweight I-shaped joists** consist of a top and bottom chords of laminated or solid construction separated by OSB or plywood. They are used in light commercial and residential construction and can span a longer distance than joists.

e) **Trusses** are standard sized wood member connected with metal plates. Their typical depths ranges from 12" to 36", and they can span between 24' to 40'.

b. Steel

a) **Beam and girder system**
Girders span longer distances, from 25' to 40', while beams span shorter distances, from 8' to 10'. Steel decking spans between the beams. A concrete topping is often poured over the decking to form the floor slab.

b) **Open-web steel joists** can span up to 60'. Long span joists can span up to 96', and deep long span joists can span up to 144'.

c. Concrete

a) **Cast-in-place** concrete structures require formwork and take longer to build. They include **one-way systems** and **two-way systems**. Common one-way systems include the *beam-and-girder* system and *concrete joist* system. Common two-way systems include *flat slab*, *flat plate,* and *waffle slab.*

Flat plates are 6" to 12" thick, designed to span in both directions directly into columns, limited to light loads, and can span up to 25'.

When live loads are heavier or the spans are large, the slab thickness increases around the columns (drop panels) to provide greater resistance against shear. Flat plates become **flat slabs**. Sometimes column caps are also used to resist shear. Flat slabs can span up to 30'.

Waffle slabs use reusable fiberglass or metal forms. They can provide longer spans of up to 40', and support heavier loads than a flat slab.

b) **Precast** components are formed in the factory, have better quality, and can be erected faster.

c) **Post-tension** concrete utilizes the advantages of steel and concrete. The post-tension steel or **tendons** are stressed after the concrete is poured and cured. They are useful at locations where high-strength is required and precast members are hard to be transported to the site.

d. Masonry

Masonry is strong in compression, but weak in tension and bending. There are three types of masonry walls:

a) **Single wythe walls:** A **wythe** is a continuous vertical section of masonry one unit in thickness.

b) **Double wythe walls:** A masonry wall composed of two walls with a space between them.

c) **Cavity walls:** It is composed of two masonry walls separated by an air space. The outer wall faces the outside of the building structure and is made of brick. The inner wall may be masonry units such as concrete block, structural clay, brick, or reinforced concrete.

Double wythe walls and cavity walls may be reinforced, grouted, or ungrouted.

Slender ratio is the wall's unsupported height to its thickness.

e. Composite construction

Composite construction is any structural system using two or more materials to act together, such as reinforced concrete, open-web steel joists with wood chords, and composite steel decking with concrete topping.

2) Complex structural systems

a. Arches are often selected for their aesthetic appeal, and can be constructed of steel, concrete, wood, stone, brick, etc. They can span from 50' to 240' for wood, 20' to 320' for concrete, or 50' to 500' for steel.

b. Folded plates are assemblies of flat plates inclined in different directions and joined along their longitudinal edges. They can span 30' to 100'.

c. **Inflatable structures** can only resist loads in tension, and are held in place by constant air pressure that is higher than the outside air pressure. **Double-skin inflatable structures** are created by inflation of the void between the double skins, and eliminate the need for air locks at entrances and exits.

d. **Rigid frames** are built so that vertical and horizontal members work together as a unit. They are the load-resisting skeletons built with curved or straight members interconnected by mostly rigid connections, which resist movements induced at the joints of members. Their members can take shear, bending moment, and axial loads.

e. **Space frames** are rigid, lightweight, truss-like structures built from interlocking struts in a geometric pattern. They can be used to span large areas with few interior supports. Their span to depth ratio may be from 20:1 to 30:1.

f. **Stressed-skin structures** include panels of a sheathing materials attached on one or both sides of intermediate web members in a way that the panels acts as the I-beam, with the intermediate members being the webs and the sheathing being the flange. They are often made of wood, and can span from 12' to 35'.

g. **Suspension structures** are supported by cables (hangers) hung from the parabolic sag of high-strength, large steel cables. They are seen in suspension bridges, or large roofs of stadiums or other spaces that require large spans.

h. **Thin-shell structures** are lightweight constructions utilizing shell elements. These elements, typically curved, are assembled to make large structures, such as aircraft fuselages, boat hulls, airport terminals, and the roofs of other large buildings.

   The main thin-shell structures are:
   a) **Concrete shell structures,** often cast as a monolithic dome or saddle roof or stressed ribbon bridge.
   b) **Lattice shell structures**, also called gridshell structures, often in the form of a hyperboloid structure or a geodesic dome.
   c) **Membrane structures**, which include fabric structures and other tensile structures, pneumatic structures, and cable domes.

i. **Trusses** are an assembly of beams or other elements that create a rigid structure. There are many different kinds of trusses. See the link below:
   https://en.wikipedia.org/wiki/Truss

3) **Selection criteria**

a. **Building function and occupancy** is a main factor in determining a structural system. An arena or an airport terminal needs a huge span. An office works well with spans between 30' to 40'.

   A developer may want to have as many floors as possible within a limited building height, which may require a concrete flat-plate with columns spaced closely. A building requiring large mechanical ducts may require deep-span, open-web trusses. A lab housing delicate, motion-sensitive equipment may require a massive concrete structure.

b. **Cost**

   Structural systems are often not visible, and can be the first in line when a project requires cost-cutting or value engineering.

   The first way to cut cost is to select the proper system. For example, wood trusses may cost less than steel trusses. A hybrid system of steel trusses with a wood deck may cost less than steel trusses with metal deck.

   The second way to cut cost is to fine tune a structural system. For example, spanning the girders across the smaller dimension of a space may reduce the girder size and save money. Changing the spacing of the beams may also save money.

c. **Feasibility**
Construction time is an important factor affecting the feasibility of a structural system. Some areas may have a very short construction season because of the local climate and weather. The cost of financing may require a very short construction time, and will make a tilt-up construction more feasible than a masonry building.

High costs and availability of skilled workers as well as availability of heavy construction equipment may make a complicated structural system unfeasible.

d. **Fire resistance**
The entire building, especially the structural systems must meet fire resistance code requirements.

First, the combustibility of the structural members must be considered. Second, the possibilities of the loss of strength of structural members during intense heat. For example, steel will not burn but will bend and collapse during intense heat, and must be protected with fire proofing.

Concrete and masonry are fire resistant and are not substantially weakened during intense heat. A steel framing system may be cheaper than a concrete framing, but may be more expensive in the long run after adding the cost of fireproofing.

e. **Integration with other building systems**
A structural system must integrate with architectural, mechanical, electrical, and plumbing systems. For example, bearing walls may need to have veneers, ductwork and pipes that run around a structural system, and electrical wires that run among it. Tilt-up concrete may require furred-out columns and pre-tensioned concrete floors may present a challenge for adding underground plumbing lines during a remodel.

f. **Loads**
This is the primary function for structural systems. They need to resist dead loads, live loads, vertical loads, and lateral forces.

For high-rise buildings, loads can determine the look of the buildings, such as the stepping Sears Towers, and the Bank of China Building in Hong Kong with diagonal framing.

g. **Social and cultural influences**
Social and cultural influences are important in selecting structural systems. A tilt-up concrete panel building may be appropriate in an industrial park, while a masonry bearing wall system may be suitable for a historical area with similar buildings.

h. Style

An international style building may dictate a concrete structure or a steel and glass structural system. A colonial style neighborhood may dictate a simple stud framing system.

4. **Mechanical systems**

Mechanical systems refer to heating, ventilation, and air conditioning (HVAC). Heating and ventilation can be achieved by active or passive systems. Passive systems utilize natural winds and sun to achieve heating and ventilation. Active systems use machines to achieve the same goal. For example, a supermarket often requires a heavy air handler that is normally located on the mezzanine floor in the back, and condensers on the rooftop. This heavy HVAC equipment has a big impact on the building design.

HVAC systems also require vertical chases or shafts and concealed ceiling spaces for ductwork. The mechanical room should be hidden from public views, and also minimize the length of the ductwork.

5. **Electrical systems**

An architect need to coordinate with an electrical engineer and know the **point of connection (POC)** and size and location of major conduits for electrical wires and the size of electrical service for a project. For example, in a residential home, the panel size is usually 100 amp or 200 amp. Some old homes still have 100-amp service, but most new homes use at least a 200-amp electrical panel.

An architect needs to coordinate with the electrical engineer regarding the electrical switchgear clearance requirements (3 ft. clearance in front of the switchgear) and electrical room location. Electrical rooms typically have heavy equipment, and should be placed at ground floor if possible.

Pipe guards/bollards are necessary for the protection of gas meters, electrical transformers, switchgears, trash enclosures, building corners, etc., if they are exposed to vehicle traffic.

Many people are not sure where to use **ground fault circuit interrupter (GFCI)** outlets. Here is some useful information:

GFCI is used in areas with water or moisture to prevent electric shock and save lives, such as the bathroom, kitchen, laundry room, or in the garage of a home.

There's a GFCI rule that says any outlet within 6 ft. of a sink needs a GFCI outlet.

"When measuring distances for GFCI protection, the distance is the shortest path that a cord would take, without piercing a floor, wall, ceiling, or fixed barrier; or passing through a door, doorway, or window."

This new measurement clarification is a much-needed addition to the NEC GFCI rules as it clarifies how to properly handle rules that include measurements. For example, there's a GFCI rule that says any outlet within 6 ft. of a sink needs a GFCI outlet. This type of regulation created confusion as to what those six feet really means. For example, an outlet that is on the other side of the wall from a sink – not even in the same room. Does that qualify as within six feet of the sink? That type of confusion is gone with this specific guideline that specifies what those measurements really mean.

There is, however, one key exception based on the wording of the code. If an outlet is under a sink blocked by a cabinet, a cord could easily go below the sink, beyond an open cabinet door and into the outlet without traveling nearly six feet. In 2014, that outlet would be considered to need the GFCI protection – and the risk is real as the outlet is directly beneath a sink and behind a cabinet that is often open. The cabinet is generally interpreted as a door, however, so the outlet wouldn't need GFCI based on the updated language of the code.
For the most part, the new details on measurement resolve questions about GFCI requirements, but they do introduce a few areas of uncertainty. Architecture, engineering, and construction professionals should check with local authorities to clarify interpretation within the jurisdictions where they work."

For source and more information, see link below:
https://www.redvector.com/articles/safety/gfci-protection-standards-to-prevent-electrocution-2017-nec-updates/

6. **Plumbing systems**
An architect needs to know the point of connection (POC) and size of major plumbing lines such as the sewer line, water line, gas line, and storm drain line for a project. These locations are coordinated with the plumbing and civil engineers.

An architect needs to coordinate the architectural, structural, mechanical, electrical, and plumbing systems, and make sure they all work together.

G. Analyze graphical representations regarding building analysis and building programming (A/E)

NCARB Requirements
You will need to understand diagrammatic graphics and evaluate how they are used to communicate building relationships and other requirements. These may include diagrams that depict horizontal and vertical circulation, space adjacencies, solar path, or views.

See discussions of bubble diagram and adjacency matrix in Chapter Five.

Appendixes

A. List of Figures

B. Official reference materials suggested by NCARB

1. Resources Available While Testing

Tips:

- *You need to read through these pages several times and become very familiar with them to save time in the real ARE exams.*

United States. American Institute of Steel Construction, Inc. *Steel Construction Manual*; Chicago, Illinois, latest edition

Beam Diagrams and Formulas:

- Simple Beam: Diagrams and Formulas - Conditions 1-3, page 3-213; Conditions 4-6, page 3-214; Conditions 7-9, page 3-215
- Beam Fixed at Both Ends: Diagrams and Formulas - Conditions 15-17, page 3-218
- Beam Overhanging One Support: Diagrams and Formulas - Conditions 24-28, pages 3-221 & 222

Dimensions and Properties:

- W Shapes 44 thru 27: Dimensions and Properties, pages 1-12 thru 17
- W Shapes 24 thru W14x145: Dimensions and Properties, pages 1-18 thru 23
- W Shapes 14x132 thru W4: Dimensions and Properties, pages 1-24 thru 29
- C Shapes: Dimensions and Properties, pages 1-36 & 37
- Angles: Properties, pages 1-42 thru 49
- Rectangular HSS: Dimensions and Properties, pages 1-74 thru 91
- Square HSS: Dimensions and Properties, pages 1-92 thru 95
- Round HSS: Dimensions and Properties, pages 1-96 thru 100

United States. International Code Council, Inc. *International Building Code***.** Country Club Hills, Illinois, latest edition.

Live and Concentrated Loads:

- Uniform and Concentrated Loads: IBC Table 1607.1, pages 340-341

2. **Typical Beam Nomenclature**

The following typical beam nomenclature is excerpted from:
United States. American Institute of Steel Construction, Inc. *Steel Construction Manual*; 14th edition. Chicago, Illinois, latest edition

E	Modulus of Elasticity of steel at 29,000 ksi	V_2	Vertical shear at right reaction point, or to left of intermediate reaction of beam, kips
I	Moment of Inertia of beam, in^4	V_3	Vertical shear at right reaction point, or to right of intermediate reaction of beam, kips
L	Total length of beam between reaction point, ft	V_x	Vertical shear at distance x from end of beam, kips
M_{max}	Maximum moment, kip-in	W	Total load on beam, kips
M_1	Maximum moment in left section of beam, kip-in	A	Measured distance along beam, in
M_2	Maximum moment in right section of beam, kip-in	B	Measured distance along beam which may be greater or less than a, in
M_3	Maximum positive moment in beam with combined end moment conditions, kip-in	L	Total length of beam between reaction points, in
M_x	Maximum at distance x from end of beam, kip-in	W	Uniformly distributed load per unit of length, kips/in
P	Concentrated load, kips	w_1	Uniformly distributed load per unit of length nearest left reaction, kips/in
P_1	Concentrated load nearest left reaction, kips	w_2	Uniformly distributed load per unit of length nearest right reaction and of different magnitude than w1, kips/in
P_2	Concentrated load nearest right reaction and of different magnitude than P_1, kips	X	Any distance measured along beam from left reaction, in
R	End beam reaction for any condition of symmetrical loading, kips	x_1	Any distance measured along overhang section of beam from nearest reaction point, in
R_1	Left end beam reaction, kips	Δ_{max}	Maximum deflection, in
R_2	Right end or intermediate beam reaction, kips	Δa	Deflection at point of load, in
R_3	Right end beam reaction, kips	Δx	Deflection at point x distance from left reaction, in
V	Maximum vertical shear for any condition of symmetrical loading, kips	Δx_1	Deflection of overhang section of beam at any distance from nearest reaction point, in
V_1	Maximum vertical shear in left section of beam, kips		

3. **Formulas Available While Testing**
   Tips:
   - *These formulas and references will be available during the real exam. You should read through them a few times before the exam to become familiar with them. This will save you a lot of time during the real exam, and will help you solve structural calculations and other problems.*

Structural:

Flexural stress at extreme fiber

$$f = \frac{Mc}{I} = \frac{M}{S}$$

Flexural stress at any fiber

$$f = \frac{My}{I}$$

where y = distance from neutral axis to fiber

Average vertical shear

$$v = \frac{V}{A} = \frac{V}{dt}$$

for beams and girders

Horizontal shearing stress at any section A-A

$$v = \frac{VQ}{Ib}$$

where Q = statical moment about the neutral axis of the entire section of that portion of the cross-section lying outside of section A-A
b = width at section A-A

Electrical

$$Foot - candles = \frac{lumens}{area\ in\ ft^2}$$

$$Foot - candles = \frac{(lamp\ lumens)\ x\ (lamps\ per\ fixture)\ x\ (number\ of\ fixtures)\ x\ (CU)\ x\ (LLF)}{area\ in\ ft^2}$$

$$Number\ of\ luminaires = \frac{(foot - candles)\ x\ (floor\ area)}{(lumens)\ x\ (CU)\ x\ (LLF)}$$

where CU = coefficient of utilization
LLF = Light Loss Factor

$$DF_{AV} = 0.2x\frac{window\ area}{floor\ area}$$

for spaces with sidelighting or toplighting with vertical monitors

$$\text{watts} = \text{volts x amperes x power factor}$$

for AC circuits only

$$Demand\ charge = maximum\ power\ demand\ x\ demand\ tariff$$

Plumbing

$$1\ psi = 2.31\ feet\ of\ water$$

$$1\ cubic\ foot = 7.5\ U.S.gallons$$

HVAC

$$\frac{BTU}{year} = peak\ heat\ loss\ x\ \frac{full - load\ hours}{year}$$

$$BTU/h = (cfm)\ x\ (1.08)\ x\ (\Delta T)$$

$$1\ kWh = 3,400\ BTU/h$$

$$1\ ton\ of\ air\ conditioning = 12,000\ BTU/h$$

$BTU/h = (U)\ x\ (A)\ x\ (T_d)$ *where Td is the difference between indoor and outdoor temperatures*

$$U = 1/R_t$$

$$U_o = \frac{(U_w \times A_w) + (U_{op} \times A_{op})}{Ao}$$

where o = total wall, w = window, and op = opaque wall

$$U_o = \frac{(U_R \times A_R) + (U_S \times A_S)}{Ao}$$

where o = total roof, R = roof, and S = skylight

$$R = x/k$$

where x = thickness of material in inches

$$Heat\,required = \frac{BTU/h}{temperature\ differential} \times (24\ hours) \times (DD\ °F)$$

where DD = degree days

Acoustics

$$\lambda = \frac{c}{f}$$

where λ = wavelength of sound (ft)
c = velocity of sound (fps)
f = frequency of sound (Hz)

$$a = SAC\ x\ S$$

where a = Absorption of a material used in space (sabins)
SAC = Sound Absorption Coefficient of the material
S = Exposed surface area of the material (ft ^{2})

$$A = \Sigma a$$

Where A = Total sound absorption of a room (sabins)
$\Sigma a = (S_1 \, x \, SAC_1) + (S_2 \, x \, SAC_2) + \ldots$

$$T = 0.05 \times \frac{V}{A}$$

where T = Reverberation time (seconds)
V = Volume of space (ft ^{3})

$$NRC = average\ SAC\ for\ frequency\ bands\ 250, 500, 1000, and\ 2000\ Hz$$

4. Common Abbreviations

Tips:

- *You need to read through these common abbreviations several times and become very familiar with them to save time in the real ARE exams.*

Professional Organizations, Societies, and Agencies

American Concrete Institute	ACI
American Institute of Architects	AIA
American Institute of Steel Construction	AISC
American National Standards Institute	ANSI
American Society for Testing and Materials	ASTM
American Society of Civil Engineers	ASCE
American Society of Heating, Refrigerating, and Air-Conditioning Engineers	ASHRAE
American Society of Mechanical Engineers	ASME
American Society of Plumbing Engineers	ASPE
Architectural Woodwork Institute	AWI
Construction Specifications Institute	CSI
Department of Housing and Urban Development	HUD
Environmental Protection Agency	EPA
Federal Emergency Management Agency	FEMA
National Fire Protection Association	NFPA
Occupational Safety and Health Administration	OSHA
U.S. Green Building Council	USGBC

Tips:

- *You need to look through the following codes and regulations & AIA contract documents several times and become very familiar with them to save time in the real ARE exams. Read some of the important sections in details.*

AIA Contract Documents

A101, Standard Form of Agreement Between Owner and Contractor - Stipulated Sum	A101
A201, General Conditions of the Contract for Construction	A201
A305, Contractor's Qualification Statement	A305
A701, Instructions to Bidders	A701
B101, Standard Form of Agreement Between Owner and Architect	B101
C401, Standard Form of Agreement Between Architect and Consultant	C401
G701, Change Order	G701
G702, Application and Certificate for Payment	G702
G703, Continuation Sheet	G703
G704, Certificate of Substantial Completion	G704

Codes and Regulations

ADA Standards for Accessible Design	ADA

International Code Council	ICC
International Building Code	IBC
International Energy Conservation Code	IECC
International Existing Building Code	IEBC
International Mechanical Code	IMC
International Plumbing Code	IPC
International Residential Code	IRC
Leadership in Energy and Environmental Design	LEED
National Electrical Code	NEC

Commonly Used Terms

Air Handling Unit	AHU
Authority Having Jurisdiction	AHJ
Building Information Modeling	BIM
Concrete Masonry Unit	CMU
Contract Administration	CA
Construction Document	CD
Dead Load	DL
Design Development	DD
Exterior Insulation and Finish System	EIFS
Furniture, Furnishings & Equipment	FF&E
Floor Area Ratio	FAR
Heating, Ventilating, and Air Conditioning	HVAC
Insulating Glass Unit	IGU
Indoor Air Quality	IAQ
Indoor Environmental Quality	IEQ
Live Load	LL
Material Safety Data Sheets	MSDS
Photovoltaic	PV
Reflected Ceiling Plan	RCP
Schematic Design	SD
Variable Air Volume	VAV
Volatile Organic Compound	VOC
British Thermal Unit	btu
Cubic Feet per Minute	cfm
Cubic Feet per Second	cfs
Cubic Foot	cu. ft. ft^3
Cubic Inch	cu. in. in^3
Cubic Yard	cu. yd. yd^3
Decibel	dB
Foot	ft
Foot-candle	fc
Gross Square Feet	gsf

Impact Insulation Class	IIC
Inch	in
Net Square Feet	nsf
Noise Reduction Coefficient	NRC
Pound	lb
Pounds per Linear Foot	plf
Pounds per Square Foot	psf
Pounds per Square Inch	psi
Sound Transmission Class	STC
Square Foot	sq. ft. sf ft^2
Square Inch	sq. in. in^2
Square Yard	sq. yd.

5. General NCARB reference materials for ARE:

Per NCARB, all candidates should become familiar with the latest version of the following codes:

International Code Council, Inc. (ICC)
International Building Code
International Mechanical Code
International Plumbing Code

National Fire Protection Association (NFPA)
Life Safety Code (NFPA 101)
National Electrical Code (NFPA 70)

National Research Council of Canada
National Building Code of Canada
National Plumbing Code of Canada
National Fire Code of Canada

American Institute of Architects
AIA Documents

6. Official NCARB reference materials matrix

Per NCARB, all candidates should become familiar with the latest version of the following:

Reference	PcM	PjM	PA	PPD	PDD	CE
ADA Standards for Accessible Design U.S. Department of Justice, Latest Edition			■	■	■	
Code of Ethics and Professional Conduct AIA Office of General Counsel. The American Institute of Architects, latest edition	■					
The Architect's Handbook of Professional Practice The American Institute of Architects John Wiley & Sons, latest edition	■	■	■	■	■	■
The Architect's Studio Companion: Rules of Thumb for Preliminary Design Edward Allen and Joseph Iano John Wiley & Sons, 6th edition, 2017			■	■	■	
Architectural Acoustics. M. David Egan. J. Ross Publishing. Reprint. Original publication McGraw Hill, latest edition				■	■	
Architectural Graphic Standards The American Institute of Architects John Wiley & Sons, latest edition			■	■	■	
Building Codes Illustrated: A Guide to Understanding the International Building Code. Francis D.K. Ching and Steven R. Winkel, FAIA, PE. John Wiley & Sons, latest edition			■	■	■	
Building Construction Illustrated Francis D. K. Ching John Wiley & Sons, latest edition				■	■	
Building Structures James Ambrose and Patrick Tripeny John Wiley & Sons, 3rd edition, Latest Edition				■	■	
CSI MasterFormat. The Construction Specifications Institute, latest edition					■	■
Daylighting Handbook I Christoph Reinhart Building Technology Press, latest edition			■	■		
Dictionary of Architecture and Construction. Cyril M. Harris. McGraw-Hill, Latest edition			■	■	■	
Framework for Design Excellence American Institute of Architects Available Online			■	■		

Reference	PcM	PjM	PA	PPD	PDD	CE
Fundamentals of Building Construction: Materials and Methods Edward Allen and Joseph Iano John Wiley & Sons, latest edition						
Green Building Illustrated Francis D.K. Ching and Ian M. Shapiro Wiley, latest edition						
The Green Studio Handbook: Environmental Strategies for Schematic Design Alison G. Kwok and Walter Grondzik Routledge, latest edition						
Heating, Cooling, Lighting: Sustainable Design Methods for Architects. Norbert Lechner. John Wiley & Sons, latest edition						
The HOK Guidebook to Sustainable Design Sandra F. Mendler, William Odell, and Mary Ann Lazarus John Wiley & Sons, latest edition						
ICC A117.1-2009 Accessible and Usable Buildings and Facilities International Code Council, 2010						
International Building Code International Code Council, latest edition						
Law for Architects: What You Need to Know. Robert F. Herrmann and the Attorneys at Menaker & Herrmann LLP. W. W. Norton, latest edition						
Legislative Guidelines and Model Law/Model Regulations National Council of Architectural Registration Boards, latest edition						
Mechanical & Electrical Equipment for Buildings. Walter T. Grondzik, Alison G. Kwok, Benjamin Stein, and John S. Reynolds, Editors. John Wiley & Sons, latest edition						
Mechanical and Electrical Systems in Buildings. Richard R. Janis and William K. Y. Tao. Prentice Hall, latest edition						
Model Rules of Conduct National Council of Architectural Registration Boards, latest edition						

Reference	PcM	PjM	PA	PPD	PDD	CE
Olin's Construction Principles, Materials, and Methods. H. Leslie Simmons. John Wiley & Sons, latest edition						
Planning and Urban Design Standards American Planning Association John Wiley & Sons, latest edition						
Plumbing, Electricity, Acoustics: Sustainable Design Methods for Architecture. Norbert Lechner. John Wiley & Sons, latest edition						
Problem Seeking: An Architectural Programming Primer William M. Peña and Steven A. Parshall John Wiley & Sons, latest edition						
Professional Practice: A Guide to Turning Designs into Buildings. Paul Segal, FAIA. W. W. Norton, latest edition						
The Professional Practice of Architectural Working Drawings. Osamu A. Wakita, Nagy R. Bakhoum, and Richard M. Linde. John Wiley & Sons, latest edition						
The Project Resource Manual: CSI Manual of Practice. The Construction Specifications Institute. McGraw-Hill, latest edition						
Simplified Engineering for Architects and Builders James Ambrose and Patrick Tripeny John Wiley & Sons, latest edition						
Site Planning and Design Handbook Thomas H. Russ McGraw-Hill, latest edition						
Space Planning Basics Mark Karlen and Rob Fleming John Wiley & Sons, latest edition						
Steel Construction Manual American Institute of Steel Construction Ingram, latest edition						
Structural Design: A Practical Guide for Architects James R. Underwood and Michele Chiuini John Wiley & Sons, latest edition						
Structures Daniel Schodek and Martin Bechthold Pearson/Prentice Hall, latest edition						

Reference	PcM	PjM	PA	PPD	PDD	CE
Sun, Wind, and Light: Architectural Design Strategies G.Z. Brown and Mark DeKay John Wiley & Sons, latest edition			■	■		
Sustainable Construction: Green Building Design and Delivery Charles J. Kibert. John Wiley & Sons, latest edition				■		
A Visual Dictionary of Architecture Francis D.K. Ching John Wiley & Sons, latest edition				■	■	

The following AIA Contract Documents have content covered in the of ARE 5.0 exams. Candidates can access them for free through their NCARB Record.

Document	PcM	PjM	PA	PPD	PDD	CE
A101-2017, Standard Form of Agreement Between Owner and Contractor where the basis of payment is a Stipulated Sum		■				■
A133-2019, Standard Form of Agreement Between Owner and Construction Manager as Constructor where the basis of payment is the Cost of the Work Plus a Fee with a Guaranteed Maximum Price		■				
A195-2008, Standard Form of Agreement Between Owner and Contractor for Integrated Project Delivery		■				
A201-2017, General Conditions of the Contract for Construction		■				■
A295-2008, General Conditions of the Contract for Integrated Project Delivery		■				
A305-1986, Contractor's Qualification Statement						■
A701-2018, Instructions to Bidders						■
B101-2017, Standard Form of Agreement Between Owner and Architect	■	■				■
B195-2008, Standard Form of Agreement Between Owner and Architect for Integrated Project Delivery		■				
C401-2017, Standard Form of Agreement Between Architect and Consultant	■	■				■

Document	PcM	PjM	PA	PPD	PDD	CE
G701-2017, Change Order						
G702-1992, Application and Certificate for Payment						
G703-1992, Continuation Sheet						
G704-2017, Certificate of Substantial Completion						

The following are some extra study materials if you have some additional time and want to learn more. If you are tight on time, you can simply look through them and focus on the sections that cover your weakness:

ACI Code 318 (Building Code Requirements for Reinforced Concrete)
American Concrete Institute, latest edition

OR
CAN/CSA-A23.1 (Concrete Materials and Methods of Concrete Construction) and CAN/CSA-A23.3 (Design of Concrete Structures for Buildings)
Canadian Standards Association

Design Value for Wood Construction
American Wood Council, latest edition

Elementary Structures for Architects and Builders, Fourth Edition
Ronald E. Shaeffer
Prentice Hall, latest edition

Introduction to Wood Design
Canadian Wood Council, latest edition

Manual of Steel Construction: Allowable Stress Design
American Institute of Steel Construction, Inc. Chicago, Illinois, latest edition

National Building Code of Canada, latest edition
Parts 1, 3, 4, 9; Appendix A
Supplement
Chapters 1, 2, 4; Commentaries A, D, F, H, I

NEHRP (National Earthquake Hazards Reduction Program) Recommended Provisions for Seismic Regulations for New Buildings and Other Structures Parts 1 and 2
FEMA latest edition

Simplified Building Design for Wind and Earthquake Forces
James Ambrose and Dimitry Vergun
John Wiley & Sons, latest edition

Simplified Design of Concrete Structures,
Eighth Edition
James Ambrose, Patrick Tripeny
John Wiley & Sons, latest edition

Simplified Design of Masonry Structures
James Ambrose
John Wiley & Sons, latest edition

Simplified Design of Steel Structures, latest edition
James Ambrose, Patrick Tripeny
John Wiley & Sons, latest edition

Simplified Design of Wood Structures, latest edition
James Ambrose
John Wiley & Sons, latest edition

Simplified Mechanics and Strength of Materials, latest edition
Harry Parker and James Ambrose
John Wiley & Sons, latest edition

Standard Specifications Load Tables &Weight Tables for Steel Joists and Joist Girders
Steel Joist Institute, latest edition

Steel Construction Manual, latest edition
American Institute of Steel Construction, latest edition

OR
Handbook of Steel Construction, Latest edition; and *CAN/CSA-S16 and CISC Commentary*
Canadian Institute of Steel Construction

Steel Deck Institute Tables
Steel Deck Institute

OR
LSD Steel Deck Tables
Caradon Metal Building Products

Structural Concepts and Systems for Architects and Engineers, latest edition
T.Y. Lin and Sidney D. Stotesbury
Van Nostrand Reinhold, latest edition

Structural Design: A Practical Guide for Architects
James Underwood and Michele Chiuini
John Wiley & Sons, latest edition

Structure in Architecture: The Building of Buildings
Mario Salvadori with Robert Heller
Prentice-Hall, latest edition

Understanding Structures
Fuller Moore
McGraw-Hill, latest edition

Wood Design Manual and *CAN/CSA-086.1 and Commentary*
Canadian Wood Council

C. Other reference materials

Chen, Gang. *Building Construction: Project Management, Construction Administration, Drawings, Specs, Detailing Tips, Schedules, Checklists, and Secrets Others Don't Tell You (Architectural Practice Simplified, 2nd edition).* ArchiteG, Inc. This book is a good introduction to architectural practice and construction documents and services, including discussions of MasterSpec format and specification sections.

Chen, Gang. ***LEED v4 Green Associate Exam Guide (LEED GA):*** *Comprehensive Study Materials, Sample Questions, Mock Exam, Green Building LEED Certification, and Sustainability*, Book 2, LEED Exam Guide series, ArchiteG.com, the latest edition. ArchiteG, Inc. Latest Edition. This is a very comprehensive and concise book on the LEED Green Associate Exam. Some readers have passed the LEED Green Associate Exam by studying this book for ten hours.

Ching, Francis. *Architecture: Form, Space, & Order.* Wiley, latest edition. It is one of the best architectural books that you can have. I still flip through it every now and then. It is a great book for inspiration.

Frampton, Kenneth. *Modern Architecture: A Critical History*. Thames and Hudson, London, latest edition. A valuable resource for architectural history.

Jarzombek, Mark M. (Author), Vikramaditya Prakash (Author), Francis D. K. Ching (Editor). *A Global History of Architecture*. Wiley, latest edition. This is a valuable and comprehensive resource for architectural history with 1000 b & w photos, 50 color photos, and 1500 b & w illustrations. It doesn't limit the topic on a Western perspective, but rather investigates through a global vision.

Trachtenberg, Marvin and Isabelle Hyman. *Architecture: From Pre-history to Post-Modernism.* Prentice Hall, Englewood Cliffs, NJ latest edition. This text is a valuable and comprehensive resource for architectural history.

D. Some Important Information about Architects and the Profession of Architecture

What Architects Do?

Architects plan and design houses, factories, office buildings, and other structures.

Duties
Architects typically do the following:

- meet with clients to determine objectives and requirements for structures
- give preliminary estimates on cost and construction time
- prepare structure specifications
- direct workers who prepare drawings and documents
- prepare scaled drawings, either with computer software or by hand
- prepare contract documents for building contractors
- manage construction contracts
- visit worksites to ensure that construction adheres to architectural plans
- seek new work by marketing and giving presentations

People need places to live, work, play, learn, shop, and eat. Architects are responsible for designing these places. They work on public or private projects and design both indoor and outdoor spaces. Architects can be commissioned to design anything from a single room to an entire complex of buildings.

Architects discuss the objectives, requirements, and budget of a project with clients. In some cases, architects provide various predesign services, such as feasibility and environmental impact studies, site selection, cost analyses, and design requirements.

Architects develop final construction plans after discussing and agreeing on the initial proposal with clients. These plans show the building's appearance and details of its construction. Accompanying these plans are drawings of the structural system; air-conditioning, heating, and ventilating systems; electrical systems; communications systems; and plumbing. Sometimes, landscape plans are included as well. In developing designs, architects must follow state and local building codes, zoning laws, fire regulations, and other ordinances, such as those requiring easy access to buildings for people who are disabled.

Computer-aided design and drafting (CADD) and building information modeling (BIM) have replaced traditional drafting paper and pencil as the most common methods for creating designs and construction drawings. However, hand-drawing skills are still required, especially during the conceptual stages of a project and when an architect is at a construction site.

As construction continues, architects may visit building sites to ensure that contractors follow the design, adhere to the schedule, use the specified materials, and meet work-quality standards. The job is not complete until all construction is finished, required tests are conducted, and construction costs are paid.

Architects may also help clients get construction bids, select contractors, and negotiate construction contracts.

Architects often collaborate with workers in related occupations, such as civil engineers, urban and regional planners, drafters, interior designers, and landscape architects.

Work Environment
Although architects usually work in an office, they must also travel to construction sites.

Architects held about 128,800 jobs in 2016. The largest employers of architects were as follows:

Architectural, engineering, and related services	68%
Self-employed workers	20%
Government	3%
Construction	2%

Architects spend much of their time in offices, where they meet with clients, develop reports and drawings, and work with other architects and engineers. They also visit construction sites to ensure clients' objectives are met and to review the progress of projects. Some architects work from home offices.

Work Schedules
Most architects work full time and many work additional hours, especially when facing deadlines. Self-employed architects may have more flexible work schedules.

How to Become an Architect
There are typically three main steps to becoming a licensed architect: completing a professional degree in architecture, gaining relevant experience through a paid internship, and passing the Architect Registration Examination.

Education
In all states, earning a professional degree in architecture is typically the first step to becoming an architect. Most architects earn their professional degree through a five-year Bachelor of Architecture degree program, intended for students with no previous architectural training. Many earn a master's degree in architecture, which can take one to five years in addition to the time spent earning a bachelor's degree. The amount of time required depends on the extent of the student's previous education and training in architecture.

A typical bachelor's degree program includes courses in architectural history and theory, building design with an emphasis on computer-aided design and drafting (CADD), structures, construction methods, professional practices, math, physical sciences, and liberal arts. Central to most architectural programs is the design studio, where students apply the skills and concepts learned in the classroom to create drawings and three-dimensional models of their designs.

Currently, thirty-four states require that architects hold a professional degree in architecture from one of the 123 schools of architecture accredited by the National Architectural Accrediting Board (NAAB). State licensing requirements can be found at the National Council of Architectural

Registration Boards (NCARB). In the states that do not have that requirement, applicants can become licensed with eight to thirteen years of related work experience in addition to a high school diploma. However, most architects in these states still obtain a professional degree in architecture.

Training

All state architectural registration boards require architecture graduates to complete a lengthy paid internship—generally three years of experience—before they may sit for the Architect Registration Examination. Most new graduates complete their training period by working at architectural firms through the Architectural Experience Program (AXP), a program run by NCARB that guides students through the internship process. Some states allow a portion of the training to occur in the offices of employers in related careers, such as engineers and general contractors. Architecture students who complete internships while still in school can count some of that time toward the three-year training period.

Interns in architectural firms may help design part of a project. They may help prepare architectural documents and drawings, build models, and prepare construction drawings on CADD. Interns may also research building codes and write specifications for building materials, installation criteria, the quality of finishes, and other related details. Licensed architects will take the documents that interns produce, make edits to them, finalize plans, and then sign and seal the documents.

Licenses, Certifications, and Registrations

All states and the District of Columbia require architects to be licensed. Licensing requirements typically include completing a professional degree in architecture, gaining relevant experience through a paid internship, and passing the Architect Registration Examination.

Most states also require some form of continuing education to keep a license, and some additional states are expected to adopt mandatory continuing education. Requirements vary by state but usually involve additional education through workshops, university classes, conferences, self-study courses, or other sources.

A growing number of architects voluntarily seek certification from NCARB. This certification makes it easier to become licensed in other states, because it is the primary requirement for reciprocity of licensing among state boards that are NCARB members. In 2014, approximately one-third of all licensed architects had the certification.

Advancement

After many years of work experience, some architects advance to become architectural and engineering managers. These managers typically coordinate the activities of employees and may work on larger construction projects.

Important Qualities

Analytical skills. Architects must understand the content of designs and the context in which they were created. For example, architects must understand the locations of mechanical systems and how those systems affect building operations.

Communication skills. Architects share their ideas, both in oral presentations and in writing, with clients, other architects, and workers who help prepare drawings. Many also give presentations to explain their ideas and designs.

Creativity. Architects design the overall look of houses, buildings, and other structures. Therefore, the final product should be attractive and functional.

Organizational skills. Architects often manage contracts. Therefore, they must keep records related to the details of a project, including total cost, materials used, and progress.

Technical skills. Architects need to use CADD technology to create plans as part of building information modeling (BIM).

Visualization skills. Architects must be able to see how the parts of a structure relate to each other. They also must be able to visualize how the overall building will look once completed.

Pay

The median annual wage for architects was $79,380 in May 2018. The median wage is the wage at which half the workers in an occupation earned more than that amount and half earned less. The lowest 10 percent earned less than $48,020, and the highest ten percent earned more than $138,120.

In May 2018, the median annual wages for architects in the top industries in which they worked were as follows:

Government	$92,940
Architectural, engineering, and related services	$78,460
Construction	$78,110

Most architects work full time and many work additional hours, especially when facing deadlines. Self-employed architects may have more flexible work hours.

Job Outlook

Employment of architects is projected to grow 4 percent from 2016 to 2026, slower than the average for all occupations.

Architects will be needed to make plans and designs for the construction and renovation of homes, offices, retail stores, and other structures. Many school districts and universities are expected to build new facilities or renovate existing ones. In addition, demand is expected for more healthcare facilities as the baby-boomer population ages and as more individuals use healthcare services. The construction of new retail establishments may also require more architects.

Demand is projected for architects with a knowledge of "green design," also called sustainable design. Sustainable design emphasizes the efficient use of resources, such as energy and water conservation; waste and pollution reduction; and environmentally friendly design, specifications, and materials. Rising energy costs and increased concern about the environment have led to many new buildings being built with more sustainable designs.

The use of CADD and, more recently, BIM, has made architects more productive. These technologies have allowed architects to do more work without the help of drafters while making it easier to share the work with engineers, contractors, and clients.

Job Prospects

With a high number of students graduating with degrees in architecture, very strong competition for internships and jobs is expected. Competition for jobs will be especially strong at the most prestigious architectural firms. Those with up-to-date technical skills—including a strong grasp of CADD and BIM—and experience in sustainable design will have an advantage.

Employment of architects is strongly tied to the activity of the construction industry. Therefore, these workers may experience periods of unemployment when there is a slowdown in requests for new projects or when the overall level of construction falls.

State & Area Data

Occupational Employment Statistics (OES)

The Occupational Employment Statistics (OES) program produces employment and wage estimates annually for over 800 occupations. These estimates are available for the nation as a whole, for individual states, and for metropolitan and nonmetropolitan areas. The link below goes to OES data maps for employment and wages by state and area.
https://www.bls.gov/oes/current/oes171011.htm#st

Projections Central

Occupational employment projections are developed for all states by Labor Market Information (LMI) or individual state Employment Projections offices. All state projections data are available at www.projectionscentral.com. Information on this site allows projected employment growth for an occupation to be compared among states or to be compared within one state. In addition, states may produce projections for areas; there are links to each state's websites where these data may be retrieved.

CareerOneStop

CareerOneStop includes hundreds of occupational profiles with data available by state and metro area. There are links in the left-hand side menu to compare occupational employment by state and occupational wages by local area or metro area. There is also a salary info tool to search for wages by zip code.

Related Occupations

Architects design buildings and related structures. Construction managers, like architects, also plan and coordinate activities concerned with the construction and maintenance of buildings and facilities. Others who engage in similar work are landscape architects, civil engineers, urban and regional planners, and designers, including interior designers, commercial and industrial designers, and graphic designers.

Sources of Additional Information

Disclaimer:
Links to non-BLS Internet sites are provided for your convenience and do not constitute an endorsement.

Information about education and careers in architecture can be obtained from:

- The American Institute of Architects, 1735 New York Ave. NW., Washington, DC 20006. Internet: http://www.aia.org
- National Architectural Accrediting Board: http://www.naab.org/
- National Council of Architectural Registration Boards, Suite 1100K, 1801 K St. NW., Washington, D.C. 20006. Internet: http://www.ncarb.org
  OOH ONET Codes 17-1011.00"

Source: Bureau of Labor Statistics, U.S. Department of Labor, *Occupational Outlook Handbook*, Architects, on the Internet at https://www.bls.gov/ooh/architecture-and-engineering/architects.htm (visited June 06, 2019).

Last Modified Date: Friday, April 12, 2019

Note:
Please check the website above for the latest information.

E. AIA Compensation Survey

Every 3 years, AIA publishes a Compensation Survey for various positions at architectural firms across the country. It is a good idea to find out the salary before you make the final decision to become an architect. If you are already an architect, it is also a good idea to determine if you are underpaid or overpaid.

See following link for some sample pages for the latest AIA Compensation Survey:

https://www.aia.org/resources/8066-aia-compensation-report

F. So ... You would Like to Study Architecture

To study architecture, you need to learn how to draft, how to understand and organize spaces and the interactions between interior and exterior spaces, how to do design, and how to communicate effectively. You also need to understand the history of architecture.

As an architect, a leader for a team of various design professionals, you not only need to know architecture, but also need to understand enough of your consultants' work to be able to coordinate them. Your consultants include soils and civil engineers, landscape architects, structural, electrical, mechanical, and plumbing engineers, interior designers, sign consultants, etc.

There are two major career paths for you in architecture: practice as an architect or teach in colleges or universities. The earlier you determine which path you are going to take, the more likely you will be successful at an early age. Some famous and well-respected architects, like my USC alumnus Frank Gehry, have combined the two paths successfully. They teach at universities and have their own architectural practices. Even as a college or university professor, people respect you more if you have actual working experience and have some built projects. If you only teach in colleges or universities but have no actual working experience and have no built projects, people will consider you as a "paper" architect, and they are not likely to take you seriously, because they will think you probably do not know how to put a real building together.

In the United States., if you want to practice architecture, you need to obtain an architect's license. It requires a combination of passing scores on the Architectural Registration Exam (ARE) and eight years of education and/or qualified working experience, including at least one year of working experience in the United States. Your working experience needs to be under the supervision of a licensed architect to be counted as qualified working experience for your architect's license.

If you work for a landscape architect or civil engineer or structural engineer, some state architectural licensing boards will count your experience at a discounted rate for the qualification of your architect's license. For example, two years of experience working for a civil engineer may be counted as one year of qualified experience for your architect's license. You need to contact your state's architectural licensing board for specific licensing requirements for your state.

If you want to teach in colleges or universities, you probably want to obtain a master's degree or a Ph.D. It is not very common for people in the architectural field to have a Ph.D. One reason is that there are few Ph.D. programs for architecture. Another reason is that architecture is considered a profession and requires a license. Many people think an architect's license is more important than a Ph.D. degree. In many states, you need to have an architect's license to even use the title "architect," or the terms "architectural" or "architecture" to advertise your service. You cannot call yourself an architect if you do not have an architect's license, even if you have a Ph.D. in architecture. Violation of these rules brings punishment.

To become a tenured professor, you need to have a certain number of publications and pass the evaluation for the tenure position. Publications are very important for tenure track positions. Some people say for the tenured track positions in universities and colleges, it is "publish or perish" environment.

The American Institute of Architects (AIA) is the national organization for the architectural profession. Membership is voluntary. There are different levels of AIA membership. Only licensed architects can be (full) AIA members. If you are an architectural student or an intern but not a licensed architect yet, you can join as an associate AIA member. Contact AIA for detailed information.

The National Council of Architectural Registration Boards (NCARB) is a nonprofit federation of architectural licensing boards. It has some very useful programs, such as IDP, to assist you in obtaining your architect's license. Contact NCARB for detailed information.

Back Page Promotion

You may be interested in some other books written by Gang Chen:

A. **ARE Mock Exam series & ARE Exam Guide series. See the following link:**
http://www.GreenExamEducation.com

B. **LEED Exam Guides series. See the following link:**
http://www.GreenExamEducation.com

C. ***Building Construction:*** *Project Management, Construction Administration, Drawings, Specs, Detailing Tips, Schedules, Checklists, and Secrets Others Don't Tell You (Architectural Practice Simplified, 2nd edition)*
http://www.GreenExamEducation.com

D. ***Planting Design Illustrated***
http://www.GreenExamEducation.com

ARE Mock Exam Series & ARE Exam Guide Series

ARE 5.0 Mock Exam Series

Practice Management (PcM) ARE 5.0 Mock Exam (Architect Registration Examination): *ARE 5.0 Overview, Exam Prep Tips, Hotspots, Case Studies, Drag-and-Place, Solutions and Explanations.* **ISBN**: 9781612650388

Project Management (PjM) ARE 5.0 Mock Exam (Architect Registration Examination): *ARE 5.0 Overview, Exam Prep Tips, Hotspots, Case Studies, Drag-and-Place, Solutions and Explanations.* **ISBN**: 9781612650371

Programming & Analysis (PA) ARE 5.0 Mock Exam (Architect Registration Exam): *ARE 5.0 Overview, Exam Prep Tips, Hotspots, Case Studies, Drag-and-Place, Solutions and Explanations.* **ISBN**: 9781612650326

Project Planning & Design (PPD) ARE 5.0 Mock Exam (Architect Registration Examination): *ARE 5.0 Overview, Exam Prep Tips, Hotspots, Case Studies, Drag-and-Place, Solutions and Explanations.* **ISBN**: 9781612650296

Project Development & Documentation (PDD) ARE 5.0 Mock Exam (Architect Registration Examination): *ARE 5.0 Overview, Exam Prep Tips, Hotspots, Case Studies, Drag-and-Place, Solutions and Explanations*
ISBN: 9781612650258

Construction & Evaluation (CE) ARE 5.0 Mock Exam (Architect Registration Examination): *ARE 5.0 Overview, Exam Prep Tips, Hotspots, Case Studies, Drag-and-Place, Solutions and Explanations*
ISBN: 9781612650241

Mock California Supplemental Exam (CSE of Architect Registration Examination): *CSE Overview, Exam Prep Tips, General Section and Project Scenario Section, Questions, Solutions and Explanations.* **ISBN**: 9781612650159

ARE 5.0 Exam Guide Series

Practice Management (PcM) ARE 5.0 Exam Guide (Architect Registration Examination): *ARE 5.0 Overview, Exam Prep Tips, Guide, and Critical Content.* **ISBN**: 9781612650333

Project Management (PjM) ARE 5.0 Exam Guide (Architect Registration Examination): *ARE 5.0 Overview, Exam Prep Tips, Guide, and Critical Content.* **ISBN**: 9781612650418

Programming & Analysis (PA) ARE 5.0 Exam Guide (Architect Registration Examination): *ARE 5.0 Overview, Exam Prep Tips, Guide, and Critical Content.*
ISBN: 9781612650487

Construction and Evaluation (CE) ARE 5 Exam Guide (Architect Registration Exam):
ARE 5.0 Overview, Exam Prep Tips, Guide, and Critical Content
ISBN: 9781612650432

Other books in the ARE 5.0 Exam Guide Series are being produced. Our goal is to produce one mock exam book *plus* one guidebook for each of the ARE 5.0 exam divisions. See the following link for the latest information:
http://www.GreenExamEducation.com

LEED Exam Guides series*:* Comprehensive Study Materials, Sample Questions, Mock Exam, Building LEED Certification and Going Green

LEED (Leadership in Energy and Environmental Design) is the most important trend of development, and it is revolutionizing the construction industry. It has gained tremendous momentum and has a profound impact on our environment.

From LEED Exam Guides series, you will learn how to

1. Pass the LEED Green Associate Exam and various LEED AP + exams (each book will help you with a specific LEED exam).
2. Register and certify a building for LEED certification.
3. Understand the intent for each LEED prerequisite and credit.
4. Calculate points for a LEED credit.
5. Identify the responsible party for each prerequisite and credit.
6. Earn extra credit (exemplary performance) for LEED.
7. Implement the local codes and building standards for prerequisites and credit.
8. Receive points for categories not yet clearly defined by USGBC.

There is currently NO official book on the LEED Green Associate Exam, and most of the existing books on LEED and LEED AP are too expensive and too complicated to be practical and helpful. The pocket guides in LEED Exam Guides series fill in the blanks, demystify LEED, and uncover the tips, codes, and jargon for LEED as well as the true meaning of "going green." They will set up a solid foundation and fundamental framework of LEED for you. Each book in the LEED Exam Guides series covers every aspect of one or more specific LEED rating system(s) in plain and concise language and makes this information understandable to all people.

These pocket guides are small and easy to carry around. You can read them whenever you have a few extra minutes. They are indispensable books for all people—administrators; developers; contractors; architects; landscape architects; civil, mechanical, electrical, and plumbing engineers; interns; drafters; designers; and other design professionals.

Why is the LEED Exam Guides series needed?

A number of books are available that you can use to prepare for the LEED exams:

1. *USGBC Reference Guides*. You need to select the correct version of the *Reference Guide* for your exam.

   The *USGBC Reference Guides* are comprehensive, but they give too much information. For example, *The LEED 2009 Reference Guide for Green Building Design and Construction (BD&C)* has about 700 oversized pages. Many of the calculations in the books are too detailed for the exam. They are also expensive (approximately $200 each, so most people may not buy them for their personal use, but instead, will seek to share an office copy).

   It is good to read a reference guide from cover to cover if you have the time. The problem is not too many people have time to read the whole reference guide. Even if you do read the whole guide, you may not remember the important issues to pass the LEED exam. You need to reread the material several times before you can remember much of it.

   Reading the reference guide from cover to cover without a guidebook is a difficult and inefficient way of preparing for the LEED AP Exam, because you do NOT know what USGBC and GBCI are looking for in the exam.

2. The USGBC workshops and related handouts are concise, but they do not cover extra credits (exemplary performance). The workshops are expensive, costing approximately $450 each.

3. Various books published by a third party are available on Amazon, bn.com and books.google.com. However, most of them are not very helpful.

   There are many books on LEED, but not all are useful.

   LEED Exam Guides series will fill in the blanks and become a valuable, reliable source:

   a. They will give you more information for your money. Each of the books in the LEED Exam Guides series has more information than the related USGBC workshops.

   b. They are exam-oriented and more effective than the USGBC reference guides.

   c. They are better than most, if not all, of the other third-party books. They give you comprehensive study materials, sample questions and answers, mock exams and answers, and critical information on building LEED certification and going green. Other third-party books only give you a fraction of the information.

   d. They are comprehensive yet concise. They are small and easy to carry around. You can read them whenever you have a few extra minutes.

   e. They are great timesavers. I have highlighted the important information that you need to understand and MEMORIZE. I also make some acronyms and short sentences to help you easily remember the credit names.

> It should take you about one or two weeks of full-time study to pass each of the LEED exams. I have met people who have spent forty hours to study and passed the exams.

You can find sample texts and other information on the LEED Exam Guides series in customer discussion sections under each of my book's listing on Amazon, bn.com and books.google.com.

What others are saying about *LEED GA Exam Guide* (Book 2, LEED Exam Guide series):

"Finally! A comprehensive study tool for LEED GA Prep!

"I took the 1-day Green LEED GA course and walked away with a power point binder printed in very small print—which was missing MUCH of the required information (although I didn't know it at the time). I studied my little heart out and took the test, only to fail it by 1 point. Turns out I did NOT study all the material I needed to in order to pass the test. I found this book, read it, marked it up, retook the test, and passed it with a 95%. Look, we all know the LEED GA exam is new and the resources for study are VERY limited. This one is the VERY best out there right now. I highly recommend it."
—ConsultantVA

"Complete overview for the LEED GA exam

"I studied this book for about 3 days and passed the exam ... if you are truly interested in learning about the LEED system and green building design, this is a great place to start."
—K.A. Evans

"A Wonderful Guide for the LEED GA Exam

"After deciding to take the LEED Green Associate exam, I started to look for the best possible study materials and resources. From what I thought would be a relatively easy task, it turned into a tedious endeavor. I realized that there are vast amounts of third-party guides and handbooks. Since the official sites offer little to no help, it became clear to me that my best chance to succeed and pass this exam would be to find the most comprehensive study guide that would not only teach me the topics, but would also give me a great background and understanding of what LEED actually is. Once I stumbled upon Mr. Chen's book, all my needs were answered. This is a great study guide that will give the reader the most complete view of the LEED exam and all that it entails.

"The book is written in an easy-to-understand language and brings up great examples, tying the material to the real world. The information is presented in a coherent and logical way, which optimizes the learning process and does not go into details that will not be needed for the LEED Green Associate Exam, as many other guides do. This book stays dead on topic and keeps the reader interested in the material.

"I highly recommend this book to anyone that is considering the LEED Green Associate Exam. I learned a great deal from this guide, and I am feeling very confident about my chances for passing my upcoming exam."
—Pavel Geystrin

"Easy to read, easy to understand

"I have read through the book once and found it to be the perfect study guide for me. The author does a great job of helping you get into the right frame of mind for the content of the exam. I had started by studying the Green Building Design and Construction reference guide for LEED projects produced by the USGBC. That was the wrong approach, simply too much information with very little retention. At 636 pages in textbook format, it would have been a daunting task to get through it. Gang Chen breaks down the points, helping to minimize the amount of information but maximizing the content I was able to absorb. I plan on going through the book a few more times, and I now believe I have the right information to pass the LEED Green Associate Exam."
—Brian Hochstein

"All in one—LEED GA prep material

"Since the LEED Green Associate exam is a newer addition by USGBC, there is not much information regarding study material for this exam. When I started looking around for material, I got really confused about what material I should buy. This LEED GA guide by Gang Chen is an answer to all my worries! It is a very precise book with lots of information, like how to approach the exam, what to study and what to skip, links to online material, and tips and tricks for passing the exam. It is like the 'one stop shop' for the LEED Green Associate Exam. I think this book can also be a good reference guide for green building professionals. A must-have!"
—SwatiD

"An ESSENTIAL LEED GA Exam Reference Guide

"This book is an invaluable tool in preparation for the LEED Green Associate (GA) Exam. As a practicing professional in the consulting realm, I found this book to be all-inclusive of the preparatory material needed for sitting the exam. The information provides clarity to the fundamental and advanced concepts of what LEED aims to achieve. A tremendous benefit is the connectivity of the concepts with real-world applications.

"The author, Gang Chen, provides a vast amount of knowledge in a very clear, concise, and logical media. For those that have not picked up a textbook in a while, it is very manageable to extract the needed information from this book. If you are taking the exam, do yourself a favor and purchase a copy of this great guide. Applicable fields: Civil Engineering, Architectural Design, MEP, and General Land Development."
—Edwin L. Tamang

Note:
*Other books in the **LEED Exam Guides series** are published or in the process of being produced. At least **one book will eventually be produced for each of the LEED exams.** The series include:*

***LEED v4 Green Associate Exam Guide (LEED GA):** Comprehensive Study Materials, Sample Questions, Mock Exam, Green Building LEED Certification, and Sustainability*, LEED Exam Guide series, ArchiteG.com. Latest Edition.

LEED GA MOCK EXAMS (LEED v4): *Questions, Answers, and Explanations: A Must-Have for the LEED Green Associate Exam, Green Building LEED Certification, and Sustainability*, LEED Exam Guide series, ArchiteG.com. Latest Edition

LEED v4 BD&C EXAM GUIDE: *A Must-Have for the LEED AP BD+C Exam: Comprehensive Study Materials, Sample Questions, Mock Exam, Green Building Design and Construction, LEED Certification, and Sustainability*, LEED Exam Guide series, ArchiteG.com. Latest Edition.

LEED v4 BD&C MOCK EXAMS: *Questions, Answers, and Explanations: A Must-Have for the LEED AP BD+C Exam, Green Building LEED Certification, and Sustainability*, LEED Exam Guide series, ArchiteG.com. Latest Edition.

LEED v4 ID&C Exam Guide: *A Must-Have for the LEED AP ID+C Exam: Study Materials, Sample Questions, Green Interior Design and Construction, Green Building LEED Certification, and Sustainability,* LEED Exam Guide series, ArchiteG.com. Latest Edition.

LEED v4 AP ID+C MOCK EXAM: *Questions, Answers, and Explanations: A Must-Have for the LEED AP ID+C Exam, Green Building LEED Certification, and Sustainability.* LEED Exam Guide series, ArchiteG.com. Latest Edition.

LEED v4 AP O+M MOCK EXAM: Questions, Answers, and Explanations: *A Must-Have for the LEED AP O+M Exam, Green Building LEED Certification, and Sustainability.* LEED Exam Guide series, ArchiteG.com. Latest Edition.

LEED v4 O&M EXAM GUIDE: *A Must-Have for the LEED AP O+M Exam: Comprehensive Study Materials, Sample Questions, Mock Exam, Green Building Operations and Maintenance, LEED Certification, and Sustainability*, LEED Exam Guide series, ArchiteG.com. Latest Edition.

LEED v4 HOMES EXAM GUIDE: *A Must-Have for the LEED AP Homes Exam: Comprehensive Study Materials, Sample Questions, Mock Exam, Green Building LEED Certification, and Sustainability*, LEED Exam Guide series, ArchiteG.com. Latest Edition.

LEED v4 ND EXAM GUIDE: *A Must-Have for the LEED AP Neighborhood Development Exam: Comprehensive Study Materials, Sample Questions, Mock Exam, Green Building LEED Certification, and Sustainability*, LEED Exam Guide series, ArchiteG.com. Latest Edition.

How to order these books:
You can order the books listed above at:
http://www.GreenExamEducation.com

OR
http://www.ArchiteG.com

Building Construction

Project Management, Construction Administration, Drawings, Specs, Detailing Tips, Schedules, Checklists, and Secrets Others Don't Tell You (Architectural Practice Simplified, 2nd edition)

Learn the Tips, Become One of Those Who Know Building Construction and Architectural Practice, and Thrive!

For architectural practice and building design and construction industry, there are two kinds of people: those who know, and those who don't. The tips of building design and construction and project management have been undercover—until now.

Most of the existing books on building construction and architectural practice are too expensive, too complicated, and too long to be practical and helpful. This book simplifies the process to make it easier to understand and uncovers the tips of building design and construction and project management. It sets up a solid foundation and fundamental framework for this field. It covers every aspect of building construction and architectural practice in plain and concise language and introduces it to all people. Through practical case studies, it demonstrates the efficient and proper ways to handle various issues and problems in architectural practice and building design and construction industry.

It is for ordinary people and aspiring young architects as well as seasoned professionals in the construction industry. For ordinary people, it uncovers the tips of building construction; for aspiring architects, it works as a construction industry survival guide and a guidebook to shorten the process in mastering architectural practice and climbing up the professional ladder; for seasoned architects, it has many checklists to refresh their memory. It is an indispensable reference book for ordinary people, architectural students, interns, drafters, designers, seasoned architects, engineers, construction administrators, superintendents, construction managers, contractors, and developers.

You will learn:

1. How to develop your business and work with your client.
2. The entire process of building design and construction, including programming, entitlement, schematic design, design development, construction documents, bidding, and construction administration.
3. How to coordinate with governing agencies, including a county's health department and a city's planning, building, fire, public works departments, etc.
4. How to coordinate with your consultants, including soils, civil, structural, electrical, mechanical, plumbing engineers, landscape architects, etc.
5. How to create and use your own checklists to do quality control of your construction documents.
6. How to use various logs (i.e., RFI log, submittal log, field visit log, etc.) and lists (contact list, document control list, distribution list, etc.) to organize and simplify your work.
7. How to respond to RFI, issue CCDs, review change orders, submittals, etc.
8. How to make your architectural practice a profitable and successful business.

Planting Design Illustrated

A Must-Have for Landscape Architecture: A Holistic Garden Design Guide with Architectural and Horticultural Insight, and Ideas from Famous Gardens in Major Civilizations

One of the most significant books on landscaping!

This is one of the most comprehensive books on planting design. It fills in the blanks of the field and introduces poetry, painting, and symbolism into planting design. It covers in detail the two major systems of planting design: formal planting design and naturalistic planting design. It has numerous line drawings and photos to illustrate the planting design concepts and principles. Through in-depth discussions of historical precedents and practical case studies, it uncovers the fundamental design principles and concepts, as well as the underpinning philosophy for planting design. It is an indispensable reference book for landscape architecture students, designers, architects, urban planners, and ordinary garden lovers.

What Others Are Saying about *Planting Design Illustrated* …

"I found this book to be absolutely fascinating. You will need to concentrate while reading it, but the effort will be well worth your time."
—Bobbie Schwartz, former president of APLD (Association of Professional Landscape Designers) and author of *The Design Puzzle: Putting the Pieces Together*.

"This is a book that you have to read, and it is more than well worth your time. Gang Chen takes you well beyond what you will learn in other books about basic principles like color, texture, and mass."
—Jane Berger, editor & publisher of gardendesignonline

"As a longtime consumer of gardening books, I am impressed with Gang Chen's inclusion of new information on planting design theory for Chinese and Japanese gardens. Many gardening books discuss the beauty of Japanese gardens, and a few discuss the unique charms of Chinese gardens, but this one explains how Japanese and Chinese history, as well as geography and artistic traditions, bear on the development of each country's style. The material on traditional Western garden planting is thorough and inspiring, too. *Planting Design Illustrated* definitely rewards repeated reading and study. Any garden designer will read it with profit."
—Jan Whitner, editor of the *Washington Park Arboretum Bulletin*

"Enhanced with an annotated bibliography and informative appendices, *Planting Design Illustrated* offers an especially "reader friendly" and practical guide that makes it a very strongly recommended addition to personal, professional, academic, and community library gardening & landscaping reference collection and supplemental reading list."
—Midwest Book Review

"Where to start? *Planting Design Illustrated* is, above all, fascinating and refreshing! Not something the lay reader encounters every day, the book presents an unlikely topic in an easily digestible, easy-to-follow way. It is superbly organized with a comprehensive table of contents, bibliography, and appendices. The writing, though expertly informative, maintains its accessibility throughout and is a joy to read. The detailed and beautiful illustrations expanding on the concepts presented were my favorite portion. One of the finest books I've encountered in this contest in the past 5 years."
—Writer's Digest 16th Annual International Self-Published Book Awards Judge's Commentary

"The work in my view has incredible application to planting design generally and a system approach to what is a very difficult subject to teach, at least in my experience. Also featured is a very beautiful philosophy of garden design principles bordering poetry. It's my strong conviction that this work needs to see the light of day by being published for the use of professionals, students & garden enthusiasts."
—Donald C. Brinkerhoff, FASLA, chairman and CEO of Lifescapes International, Inc.

Index

www.ingramcontent.com/pod-product-compliance
Lightning Source LLC
LaVergne TN
LVHW080042210926
848481LV00011B/3257
9781612650487